Synthesis Lectures on Human Language Technologies

Series Editor

Graeme Hirst, Department of Computer Science, University of Toronto, Toronto, Canada

The series publishes topics relating to natural language processing, computational linguistics, information retrieval, and spoken language understanding. Emphasis is on important new techniques, on new applications, and on topics that combine two or more HLT subfields.

Inderjeet Mani

Narrative and Generative AI

A Computational Account

Second Edition

 Springer

Inderjeet Mani
Hua Hin, Thailand

ISSN 1947-4040 ISSN 1947-4059 (electronic)
Synthesis Lectures on Human Language Technologies
ISBN 978-3-031-94057-6 ISBN 978-3-031-94058-3 (eBook)
https://doi.org/10.1007/978-3-031-94058-3

1st edition: © Morgan & Claypool Publishers 2013
2nd edition: © The Editor(s) (if applicable) and The Author(s), under exclusive license to Springer Nature Switzerland AG 2026

This work is subject to copyright. All rights are solely and exclusively licensed by the Publisher, whether the whole or part of the material is concerned, specifically the rights of translation, reprinting, reuse of illustrations, recitation, broadcasting, reproduction on microfilms or in any other physical way, and transmission or information storage and retrieval, electronic adaptation, computer software, or by similar or dissimilar methodology now known or hereafter developed.

The use of general descriptive names, registered names, trademarks, service marks, etc. in this publication does not imply, even in the absence of a specific statement, that such names are exempt from the relevant protective laws and regulations and therefore free for general use.

The publisher, the authors and the editors are safe to assume that the advice and information in this book are believed to be true and accurate at the date of publication. Neither the publisher nor the authors or the editors give a warranty, expressed or implied, with respect to the material contained herein or for any errors or omissions that may have been made. The publisher remains neutral with regard to jurisdictional claims in published maps and institutional affiliations.

This Springer imprint is published by the registered company Springer Nature Switzerland AG
The registered company address is: Gewerbestrasse 11, 6330 Cham, Switzerland

If disposing of this product, please recycle the paper.

Preface

The advent of Generative AI and Large Language Models (LLMs) has led to surprisingly capable systems such as ChatGPT and its successors. These AI tools have already revolutionized aspects of the practice of science and medicine, and have shown astonishing promise in text and computer code generation, image analysis, and certain problem-solving tasks. The public at large, including students and creative writers, have started using them as design and writing aids, as well as to summarize and analyze texts, sometimes assigning machines tasks that in a human would require substantial knowledge and inference. However, today's AI has obvious failings, which many users are all too aware of, and it is already posing enormous challenges to society.

One key topic when it comes to artificial cognitive capabilities is understanding and creating stories. Is it only a matter of time before such systems will be able to wholeheartedly assist or emulate—and possibly replace—human readers and writers? Though AI can mimic and even improve upon certain kinds of human performance, aren't the bots fundamentally too inaccurate? Aren't they crippled by their obvious deficits in commonsense and other types of reasoning, and of course their lack of culture and embodiment?

Until now, there has been no technical book that takes an in-depth look at the scope and limitations of today's Generative AI tools in terms of narrative understanding and generation. This book discusses the key problems, approaches, and outstanding challenges in modeling the narrative structure of stories. It introduces classical narratological concepts from literary theory and their mapping to computational approaches from Natural Language Processing (NLP) and AI. Crucially, it addresses the issue of to what extent current AI systems can model, even implicitly, cognitive aspects of stories that we are all familiar with.

These aspects include the underlying events of the story, their time lines and the characters involved, along with their beliefs, goals and motivations, the impacts of their actions, and inferences as to other characters' motives and reactions. They also include the overall plot, the trajectories and relationships of the characters, and an audience's responses to both the evolving story line as well as the plausibility of the setting, events, and characters.

The book's main focus is the extent to which current AI can automate the above cognitive aspects of both story understanding and generation, and the outstanding challenges that need to be addressed.

This book builds upon my 2013 monograph *Computational Modeling of Narrative* [1]. That book aimed to provide a systematic foundation integrating together narratology, AI, and computational linguistics and NLP. Given the meteoric rise of AI since then, the new book substantially rewrites and extends the earlier work, providing a more in-depth and yet accessible introduction to topics like Information Theory, Deep Learning, LLMs, Transformers, and other aspects of the current AI revolution. It also provides a series of entirely new tests that probe deep into AI abilities to understand fundamental as well as rather subtle aspects of narrative. Such capabilities can be used to answer questions about narratives and rewrite or generate short narratives.

This book can serve as a textbook, a classroom supplement, or as an additional resource. It can be used in a one-semester senior undergraduate course or graduate seminar. The audience for this book is primarily computer scientists and literary scholars, as well as linguists, digital humanities researchers, data scientists, and others interested in narrative. Last but not least, I include creative writers, including those who seek to experiment on fiction using Gen AI and of course programmers of story generation systems.

Aiming at such a wide audience may seem overly ambitious, so a few words of explanation are in order.

As someone who has published scholarly works in computer science as well as humanities narratology, I am aware of the challenges and pitfalls in communicating across disciplines. Computer scientists, including AI researchers, data scientists, and game developers, are often unaware of the relevance of narratological principles that have come from research in the humanities. It is too much to ask them to become familiar with relevant literary theories and the styles of communication that inform those disciplines.

Likewise, humanities scholars and students are often limited by presentations that seem overwhelming or even incomprehensible in their computational and mathematical details—leaving it unclear what the scope of AI is for their fields.

Finally, many fields, including linguistics, are quite technical but sometimes don't require a college-level math foundation.

To bridge such gaps effectively, one needs to avail of pedagogical strategies that emphasize the important intuitions that underlie approaches in these different fields. At the same time, one must provide enough technical content for readers, including students, who lack a sufficient foundation to read the current computational literature.

As a result, I have made the text throughout as user-friendly as possible, with plenty of examples and use of metaphors. I have also confined the math and computational details to starred sections, each of which is preceded by a section that explains the relevant intuitions without any math. (I have also added a thin bar to the left of each para within each starred section, so if the section heading isn't visible, the reader will still be guided as to whether to skip that para). I have also checked every chapter to make sure all the non-starred sections are coherent and accessible to humanities students without any math background. Each chapter, in effect, can be read in two different ways, which each way providing standalone and complete accounts.

I have written this book mainly to share my own discoveries, and to help foster further inter-disciplinary ideas. As a writer myself, with a scholarly background in both computer science and literary theory, I have felt a strong need to bring these multiple worlds together. I continue to believe wholeheartedly in the underlying unity of science, and eventually, in a hopefully non-reductive and holistic way, of knowledge.

I would like to acknowledge the extremely valuable feedback I received on an earlier draft from three anonymous reviewers. I remain grateful to Graeme Hirst, the series editor, for his advice and steady encouragement. I also appreciate the efforts by the publisher Springer Nature in bringing this project to full fruition.

Hua Hin, Thailand
August 2025

Inderjeet Mani

Reference

1. Mani, Inderjeet (2013). *Computational Modeling of Narrative.* San Rafael, CA: Morgan & Claypool.

Contents

1 Introduction 1
   1.1 Preliminaries 1
        1.1.1 Chapter Overview 3
   1.2 What is a Narrative? 4
        1.2.1 Definitions 4
        1.2.2 Narrative Structure 5
   1.3 Information Theory and Language Modeling 7
        1.3.1 Fundamentals 7
        1.3.2 The Original Language Model 10
        1.3.3 Entropy 11
        1.3.4 Cross-Entropy 14
        1.3.5 Probability of a Sentence 16
        1.3.6 LLM Overview 18
   1.4 Meaning 20
        1.4.1 What is Meaning? 20
        1.4.2 Distributional Semantics 22
        1.4.3 Do Machines Really Understand? 23
   1.5 Deep Models and Human Linguistic Abilities 24
        1.5.1 Biological Validity of Neural Models 25
        1.5.2 Deep Learning Versus Human Language Acquisition 25
        1.5.3 Language Universals and Information Theory 26
        1.5.4 Efficiency in Information Theory and NL 26
   1.6 AI Limitations 28
        1.6.1 LLM Limitations 28
        1.6.2 Problems not Specific to LLMs 30
        1.6.3 Social Limitations 32
        1.6.4 Limitations in Creativity 32
        1.6.5 Human Limitations 33

	1.7	Summary	33
	1.8	Book Overview	34
		1.8.1 Chapter Previews	34
		1.8.2 Protocols for Testing Examples	35
		1.8.3 NarrativeML	35
	References		36
2	**Deep Learning and Transformers**		**41**
	2.1	Introduction	41
	2.2	Simple Linear Perceptrons	42
		2.2.1 SLP Architecture	42
		2.2.2 Training a SLP	45
		2.2.3 SLP Limitations	47
	2.3	Multi-layer Neural Nets	48
		2.3.1 FF NN Architecture*	49
		2.3.2 FF NN Example	51
		2.3.3 FF NN Example: Computational Details*	53
		2.3.4 What's Wrong with Tiny Tot?	56
		2.3.5 Training a FF NN	56
	2.4	Word Embeddings	60
		2.4.1 Word2vec	60
		2.4.2 GloVe	62
		2.4.3 GPT Embeddings	63
	2.5	Self-attention	64
		2.5.1 Background	64
		2.5.2 Computing Self-attention*	65
		2.5.3 A Detailed Example	66
		2.5.4 Handling Richer Context*	68
		2.5.5 Multi-head Attention	69
		2.5.6 Summary	70
	2.6	Transformers	70
		2.6.1 Introduction	71
		2.6.2 Architecture*	73
		2.6.3 Language Modeling with Transformers	73
		2.6.4 Sample Specs for a Transformer*	77
		2.6.5 Bidirectional Encoders	77
	2.7	Fine-Tuning and RAG	78
		2.7.1 Fine-Tuning	78
		2.7.2 RAG	79
	2.8	Prompt Engineering	79
	2.9	Assessment and Conclusion	81
	References		83

3 Introduction to Narratology ... 85
- 3.1 Introduction ... 85
- 3.2 Narrator Characteristics ... 87
  - 3.2.1 Narrator Identity ... 87
  - 3.2.2 Narrative Distance ... 89
  - 3.2.3 Narrator Perspective: Focalization ... 91
  - 3.2.4 Focalization with LLMs ... 93
- 3.3 Narrative Levels ... 99
  - 3.3.1 Embedded Narratives and LLMs ... 99
  - 3.3.2 Narrative Threads ... 101
  - 3.3.3 Metalepsis and LLMs ... 102
  - 3.3.4 Subordinated Discourse ... 104
  - 3.3.5 Accessibility ... 105
- 3.4 Character Representations ... 107
- 3.5 Audience ... 110
  - 3.5.1 Preliminaries ... 110
  - 3.5.2 Audience Response ... 111
  - 3.5.3 Evaluating Event Outcomes ... 112
- 3.6 NarrativeML ... 115
- 3.7 Conclusion ... 118
- References ... 119

4 Plot Basics ... 123
- 4.1 Introduction ... 123
  - 4.1.1 Background ... 123
  - 4.1.2 Aristotelian Plot ... 124
  - 4.1.3 Narrative Arc ... 125
  - 4.1.4 Heroic Quests ... 127
  - 4.1.5 Narrative Functions ... 129
  - 4.1.6 Causal Models of Plot ... 134
  - 4.1.7 Comparison of Plot Models ... 143
  - 4.1.8 Narratological Implications ... 144
- 4.2 NarrativeML ... 146
- 4.3 Conclusion ... 147
- References ... 147

5 Time ... 151
- 5.1 Introduction ... 151
- 5.2 Temporal Positioning: Tense and Aspect ... 152

	5.3	Narratological Distinctions	154
		5.3.1 Narrative Time	154
		5.3.2 Narrative Order	155
	5.4	Temporal Narrative: Human Processing Results	157
	5.5	Temporal Representation	158
		5.5.1 Interval Calculus	158
		5.5.2 Temporal Annotation Scheme	160
	5.6	Automatic Temporal Analysis and Generation	166
		5.6.1 Linguistic Preliminaries	166
		5.6.2 Temporal Analysis	166
	5.7	Duration and Tempo	173
		5.7.1 Duration	173
		5.7.2 Tempo	174
	5.8	NarrativeML with Temporal Information and LLM Annotation	177
	5.9	Narratological Implications and Conclusion	180
	References		181
6	**Story Generation**		185
	6.1	Introduction	185
		6.1.1 Planning	186
	6.2	Historical Story Generation Approaches	189
		6.2.1 Character Goals	189
		6.2.2 Case-Based Reasoning	190
		6.2.3 Narrative Goals	191
		6.2.4 Planning for Interactive Narrative	193
	6.3	Temporal Generation	197
	6.4	Contemporary Narrative Generation Trends	199
		6.4.1 Plot Planning	200
		6.4.2 Evaluation Challenges	206
	6.5	NarrativeML Redux	211
	6.6	Digression: NarrativeML Plot-level Analysis	214
	6.7	NarrativeML Story Generation	214
		6.7.1 Using Noisy NarrativeML as Fabula	214
		6.7.2 Leaving Plot Out of Fabula	216
		6.7.3 Providing Partial Plot in Fabula	217
	6.8	Status of Story Generation Systems	218
	6.9	The Hand of the Creator	220
	6.10	Conclusion	220
	References		221

7	Contributions, Extensions and Future Directions		225
	7.1	Contributions	225
	7.2	Extensions	226
		7.2.1 Spatial Relations	226
		7.2.2 Image Analysis	229
	7.3	Future Directions	230
		7.3.1 Accelerating Studies in the Humanities	230
		7.3.2 Revolutionizing Co-authoring of Literature?	234
		7.3.3 Revolutionizing Co-authoring of Movies?	235
		7.3.4 Detecting Disinformation and Conspiracy Narrative Networks	236
	7.4	Final NarrativeML	236
	References		237

Appendix A: NarrativeML Annotation Guidelines 239

Index 251

Introduction

Study the science of art; Study the art of science. Learn how to see.

—Leonardo da Vinci

1.1 Preliminaries

Storytelling is one of our most fundamental and familiar intellectual and social activities, with homo sapiens often characterized as *the* storytelling animal. The habit has been with us a long time; the cave paintings at Lascaux, dating back 17,000 years, tell their animal tales using narrative techniques including sequencing in time. While the art of narrative has been practiced since antiquity, so has its study; some of the earliest aesthetic theories pertaining to narrative include those of Aristotle (384–322 BCE) and Longinus (100 or 200 CE) in Greece and Bharata (200 BCE or 200 CE) in India. Today, guidelines for narrative construction are commonplace, with narrative concepts used informally in reviews and discussions of books, movies, plays, etc.

Why do we tell stories? The evolutionary origins of narrative have given rise to considerable speculation. Based on studies of today's hunter-gatherers, it seems quite likely that narratives are evolutionary adaptations for transmitting information that can serve multiple purposes, including fostering greater co-operation in various activities such as foraging (Dubourg and Baumard 2022, Smith 2017), hunting and possibly even social control through religion.

Until recently, humans have been the principal creators of stories, with limited aid from machines. When machines were instructed to create stories, these were based on humans providing sketches of plans for the story and templates for how the story should be expressed. The results were highly restricted to specific domains. Likewise, machine understanding of natural language was limited, especially in terms of getting at meaning. Machine capabilities have, however, improved drastically, thanks to the advent of Large Language Models (**LLMs**) such as the much-discussed ones from OpenAI's GPT family. These systems are, as I will show, astounding in some of their capabilities in a way that has never been observed before in Natural Language Processing (NLP) and AI. They also seem at times severely limited and incompetent at certain tasks.

At this point, it is worthwhile to specify what I mean by AI. Historically, AI has often been defined in comparison to human intelligence; for example, a machine that performs tasks that if carried out by a human would be judged to be intelligent. Instead of comparing against humans, even though AI systems may have been originally modeled on them (and instead of falling back on hard-to-define terms like intelligence), I prefer the more practical, and very recent definition from the European Union's Artificial Intelligence Act[1]: *An 'AI system' means a machine-based system that is designed to operate with varying levels of autonomy and that may exhibit adaptiveness after deployment, and that, for explicit or implicit objectives, infers, from the input it receives, how to generate outputs such as predictions, content, recommendations, or decisions that can influence physical or virtual environments.*

This definition is broad enough to encompass historical rule-based AI systems (where humans specified the rules), some of which are discussed in this book, as well as contemporary self-learning ones (which of course properly include LLMs like GPT and such, which learn automatically from data). The definition also includes in its swathe the vast number of embedded AI systems that permeate almost all our online interactions, including search, shopping, navigation, email, and messaging. In turn, Generative AI, or Gen AI for short, *refers to computational techniques that are capable of generating seemingly new, meaningful content such as text, images, or audio from training data* (Feuerriegel et al. 2023). Note that the text being generated by the Gen AI system could be an answer, a caption, a summary, a story, or even a literary analysis.

What are the capabilities of these new systems for understanding and generating stories? Are there any fundamental limitations? What are the implications for human storytelling? After all, if a machine could understand stories, it could be of tremendous assistance in helping *us* understand and improve stories. As I asked a decade ago in a popular essay (Mani 2016b), might AI also limit our reliance on humans for literary criticism and book reviews? For story generation, has the era of machine novelists and AI scriptwriters finally arrived? Will human novelists and scriptwriters, along with reviewers and critics, be put out to pasture?

In this book, I will discuss these questions in the light of classical narratological concepts from literary theory and their mapping to computational approaches from natural language processing (NLP) and AI. The particular narratological terminology introduced is as close as it comes to a canonical standard for representing narratological phenomena—though it can be roundly criticized for various conceptual and empirical shortcomings. Narratology

---

[1] https://artificialintelligenceact.eu/article/3/.

1.1 Preliminaries

is also fairly systematic and formal, lending itself well in certain aspects to computational approaches.

Why do we care about classical narratological concepts instead of everyday notions of storytelling found in mass media? Here is an analogy. When listening to, performing, or composing music, understanding music theory can be very helpful. Knowing the basics of pitch, rhythm, harmony, and melody, and how this relates to acoustic information has several benefits, even though musicians and music lovers may manage without such knowledge:

- Music theory gives one a better appreciation of how the structure and organization of musical sounds can contribute to aesthetic pleasure. It thus provides a legitimate area of scholarship.
- It also provides a technical language to help musicians communicate with each other and learn instruments, as well as for teachers to impart their knowledge.
- For computational music understanding or synthesis, understanding these principles and how they relate to computational algorithms is essential.

The same points apply to narrative (using the terminology we will develop in Chap. 3). Indeed, one can argue that such knowledge is indispensable given the arrival of a potentially revolutionary new technology relevant to narrative.

1.1.1 Chapter Overview

In what follows, I will briefly define, in narratological terms, what I mean by narrative, indicating the scope of what is included in this book. I then provide an overview of the fundamental science of Information Theory, which lays the foundation for understanding the technical developments with LLMs that we will discuss in detail in Chap. 2. We introduce Zipf's Law (a power law which is reflected in lots of phenomena including narrative), and entropy, which measures how unpredictable a text is. After explaining how to calculate entropy and the probabilities of sentences, and the ideas behind language models, we discuss the predictability of different written languages like English and Chinese.

This overview then naturally leads to larger questions about meaning, where we introduce the linguistic theory of meaning that informs the LLMs and the NLP we use today, namely Distributional Semantics. The essential idea is that the meaning of a word is not some pre-specified concept or definition, but a set of relationships of different strengths arising from the word's relation to other words. We then turn to the problematic question of whether a machine actually understands the meaning of the data it's been fed, including prompts from a human. This is followed by a comparison of deep learning models and human linguistic abilities in relation to both natural language and narrative, following which we discuss the limitations of current AI approaches. After outlining the methodological approach I will follow, which involves an annotation scheme called NarrativeML, I provide a brief overview of the rest of the book.

I should mention in passing that my examples are drawn from those discussed in the previous literature, both narratological as well as computational. This book is not a study of a specific corpus or a particular historical context. Some of the narratological examples belong to a body of literary texts that have been called, somewhat contentiously, 'canonical', other examples do not. My examples have been drawn opportunistically to illustrate various points in my argument, rather than making any claims about a particular literary-historical context. Many of my experiments with LLMs, however, are applied to a corpus of crowdsourced 5-sentence stories, called the ROC corpus (Mostafazadeh and Nathanael 2016).

1.2 What is a Narrative?

1.2.1 Definitions

With **narrative**, it's best to begin with the everyday pre-theoretic use of the term, namely its dictionary senses as a count noun, from the Oxford English Dictionary (OED):

1. An account of a series of events, facts, etc., given in order and with the establishing of connections between them; a narration, a story, an account.
2. The part of a text, esp. a work of fiction, which represents the sequence of events, as distinguished from that dealing with dialogue, description, etc.; narration as a literary method or genre.
3. In structuralist and post-structuralist theory: a representation of a history, biography, process, etc., in which a sequence of events has been constructed into a story in accordance with a particular ideology.

The above definition is pretty close to what we mean here, though some refinements are called for. A narrative can be viewed as a form of **discourse** (in the linguistic sense of a self-contained body of text that contains one or more sentences) that involves storytelling. Dialogues of everyday life studied in the field of **discourse analysis** can include narratives, and so can histories, diaries, blogs, traditional letters, emails, instant messages, travelogues, news, political speeches, scriptures, sermons, propaganda, and scientific works, including popular science and descriptions of experiments. Storytelling can also, of course, produce more obviously literary material. Such stories can naturally be expressed in prose, and fictional works in particular can be expressed as epics, dramatic works, short stories, novels, etc.

What does not count as a narrative is equally important. An utterance such as a greeting may or may not tell a serious tale. A mention of an event (such as a bombing, love affair or war) may not count as a narrative unless it is further expanded either through elucidating its causes and/or effects, or elaborating a description of it. The classic example of Forster (1956) for what counts and does not count as a fictional plot applies here: "The king died and then the queen died" is NOT a narrative in my sense of the term, whereas "The king died and the queen died of grief" is. "He told me a story" is not a narrative unless it were further expanded, as in "He told me a story about a king who died and a queen who then died of grief." The six-word short story "For sale: Baby shoes—never worn" (attributed, perhaps apocryphally, to Hemingway) is often cited as a forerunner of today's 'flash fiction', and should certainly be considered a narrative. *Three Wise Men of Gotham* (the four-line verse) is also a narrative, while the six-line nursery rhymes *Mary Morey* and *Jack a Nory* probably aren't narratives. In short, a narrative must bottom-out (or drill-down) into more than one event, and these events must be connected logically to each other. In a conversation, only a sequence of utterances that constitutes such a coherent whole will be deemed a narrative. In a story or novel, on the other hand, the entire work is by definition a narrative. For more discussion, see Ryan (2007).

Narratives can be focused on real events, as in an eyewitness account, or on imaginary events and characters. The focus here is mainly on narrative fictions, which can involve real as well as imaginary events. There are also many aspects of narrative, including the modeling of style, subtle lexical connotations, metaphor, humor, and irony that the traditional study of narrative in the humanities (called narratology) does not concern itself with. Likewise, there is work in corpus linguistics on stylistic and narratological matters (looking mainly at lexical distributions), e.g., Herman (2005), Toolan (2009), which will not be discussed, though it complements the computational approach developed here. Finally, there is other work in Computational Literary Studies and Digital Humanities, only some of which is cited here. For more on that body of work, see Jockers and Thalken (2020) for an introduction, along with research by Rockwell and Sinclair (2022). Their collective inventory of tools, which include lexical statistics, topic modeling and named entity recognition, while useful for literary studies, is more coarse-grained compared to this book—which focuses on far more fine-grained semantic and narratological analyses.

1.2.2 Narrative Structure

I consider **Narrative Structure** to be the structure of narratives in different media. By structure, I mean representations of different phenomena that are relevant to making sense of narrative as story: these aspects include narrator perspective, narrative embeddings and threads, characters and their goals and outcomes, events, plot, time, and audience, among others. These representations more often span entire texts rather than individual sentences, and go beyond syntax and semantics to include much of what linguists broadly consider to be pragmatics.

The primary structural analysis is of the forms found in text, where **text** is a sequence of characters (i.e., written symbols) in a written work. There is a natural extension to forms found in oral storytelling, and storytelling across **media**, which I intend broadly to include literature, dance, film, theater, puppetry, animation, cave paintings, games, etc. However, I will be confining the discussion in this book to narratives from literature, with occasional passing references to film.[2]

---

[2] While poetry and drama have been discussed extensively in classical work on narratology, including Plato, Aristotle, Longinus, and Bharata, they are specialized forms that are not addressed here.

The focus on structural aspects of narrative is in itself insufficient for developing a complete theory of narrative, since structures arise in the course of storytelling from environments where agents interact with each other, involving interleaved processes of generation and understanding. Nevertheless, structure covers many of the crucial aspects of narrative that can be represented in computational terms.

Stated plainly, an agent called a **narrator** narrates a narrative to an **audience**, consisting of one or more readers, listeners, or spectators. The act of narration is an event, and a particular kind of speech act. The narrative itself can be expressed in various media, e.g., told verbally, recited, sung, danced, acted, played out, etc.

1.2.2.1 Story and Fabula

Here it is useful, especially for computation, to make the traditional narratological distinction between the underlying content of a narrative and its expression. **Story** is the content of a narrative, namely, "the chain of events (actions, happenings), plus what may be called the existents (characters, items of setting)," and **discourse** is the narrative's expression, "the means by which the content is communicated" (Chatman 1980). This distinction has been given different names in the course of narratological history: **histoire** versus **discours** in the French structuralism of Genette (1980), and **fabula** versus **sjuzhet** in the Russian formalism of Shklovsky (1973), where *fabula* is the "raw materials of the story" and *sjuzhet* is "the narrative as told or written—incorporat[ing] the procedures, emphases, and thematic devices of the literary text" (Martin 1986, pp. 107–108). (In later structuralists such as Bal (1997), the notion of fabula is stricter, so that the many versions of a given tale, e.g., *Cinderella*, or *Robin Hood*, will share the same fabula.)

This structuralist sense of *story* is also consistent with non-structuralist approaches, such as Forster's *story*, which is "a narrative of events arranged in their time-sequence" (Forster 1956). To avoid ambiguity, I will use the term *fabula* instead of *story*, except where it is obvious that the narratological sense of the latter is intended. Instead of using the term *sjuzhet*, I will use the term **discourse**, since that term is used in practically the same way in narratology and in linguistics.

1.2.2.2 Computational Implications

Fabula Versus Plot

Developers of intelligent narrative systems have to be cognizant of these distinctions. Story generation is usually decomposed into at least two stages, producing a plot, and then generating text or other media from the plot. This more or less corresponds to the decomposition found in natural language generation systems, where the first stage is deciding what to say (or content planning), and the second stage is deciding how to say it (or surface realization).[3] For narrative, the fabula is usually, but not always, instantiated in terms of the events

---

[3] Some generation researchers distinguish a sentence planning stage preparatory to surface realization.

of the entire narrative (with their participants, places, and times, and sometimes their causes and effects) in chronological and causal order prior to any verbalization thereof, with the discourse being the final generated output.

The computational *fabula* is thus usually distinct from the computational *plot*. The latter is usually at a more abstract level, covering many different stories, and thus the many variants of the story of Cinderella, with different events in them, may have the same or similar plots. Producing a fabula is usually viewed as a byproduct of planning or synthesizing a plot, while the final discourse uses text generation methods to assemble multi-sentence text. Some of the structural aspects of narratives that need to be modeled for computational purposes can be represented at the discourse level, whereas others necessarily relate to the fabula.

Causal and Other Reasoning
Given the above remarks, it should be clear that the processing of narrative therefore involves not only NLP understanding and generation but some form of causal reasoning, namely linking events in a story for both plausible inferences as to why events happen, and setting up expectations for what could happen. Since time and events are involved, it also requires temporal reasoning. Since the world inside a story involves a setting and entities that are spatially arranged and that can move, inferences about spatial relationships and motion are also involved. As we shall see, it involves reference to a narrator as well as an audience, both of whom respond to various developments in the story. This response, broadly speaking, involves emotions, and thus the ability to understand and express emotions (which from my standpoint is a kind of computation or reasoning) is also required. Are these capabilities fully met by current Generative AI systems? Though I will explore the reactions of the audience of a narrative, discussing emotion per se is precluded for reasons of space and focus.

1.3 Information Theory and Language Modeling

The focus here is on narratives in natural language. To address the applicability of LLMs to narrative, we need to review the notion of language models, which comes out of information theory as formulated in the 1940's by Shannon and Weaver (1949), which forms the theoretical basis for today's LLMs.

1.3.1 Fundamentals

In Information Theory as introduced by Shannon and Weaver (1949), a source, e.g., a speaker or author, selects a desired message from a set, expressing it in some form of information, and a transmitter changes that form into a code that is sent as a signal over a communication channel.

For example, the message could be expressed in the form of sound waves from the speaker's voice, which the transmitter could digitize into a binary code. Or the message could be an image of a face that is transmitted in some format, such as the JPEG image compression format. Naturally, the message could also be expressed in the form of text, viewed as a sequence of symbols, which the transmitter codes into some internal representation.

The receiver then converts the coded signal—which can arrive distorted by noise in the channel—back into a message (sound, pictures, text, etc.) which is provided to the destination, e.g., the hearer or user. Of course, if the internal representation used for transmission is lossy, as JPEG is, the result will not entirely recover the original message.

Bits

How much information a source expresses is related to the amount of uncertainty or freedom of choice one has when deciding which message the information conveys. Information is measured in *bits*.

To explain more about bits, we need to use some basic math—and thus the following discussion is starred. Like all starred paragraphs or sections, it can be safely skipped by those who aren't interested in the math. (However, the math in this little starred section is quite simple.) If you are concerned whether you accidentally dropped into a starred section, look to the left of the paragraph: if there's a thin vertical bar to the left of the paragraph, you are in a starred section!

> **A bit more on the bit***
>
> The total number of bits to represent a message is *the logarithm of the number of choices required to pick out the message.* The number of choices can be viewed as the number of yes/no questions that have to be asked in order to pick out the message. Thus, a signal with a code that uses three bits of information (e.g., a binary sequence 001, 010, etc.) can express eight possible messages. (We get 8 possible messages because the logarithm of 8 is 3. All logs in this book are to the base 2—unless indicated otherwise.)
>
> Yet another way of looking at it is this: if there are 8 possible yes/no decisions or choices as to which message the signal conveys, those decisions can be arrived at by simply using 3 bits of information. To make it even simpler: *n bits corresponds to 2^n choices.*

However, all symbols aren't created equal in terms of expressing information. In the case of English text, the letter 'e' is far more frequent than 'q', and the sequence 'th' more frequent than 'xp'. Knowing these differences can help save time and channel capacity when the transmitter encodes the information into a signal. Most coding systems, e.g., Morse Code, Huffman coding, etc., make good use of these differences.

1.3 Information Theory and Language Modeling

Overt Absence of Meaning

It is worth noting right away that there is no direct notion of meaning or internal structure in this system, as information content has no direct bearing on semantics. As a result, there is the complete absence of **truthfulness** of information itself: all that matters is whether the message is successfully and efficiently transmitted. This is crucial to bear in mind when discussing bias, prejudice, inauthenticity, falsehood and other aspects of information content that have always been ubiquitous in human conversation and 'messaging' (political, corporate or other), and that society is (rightfully) concerned with today and is trying (less successfully) to regulate. I will say more about this important issue in relation to storytelling in Sect. 1.5.

(Covert) Presence of Meaning

Having declared that Shannon's system has no intrinsic notion of meaning, it is worth pausing for a minute to reflect on word frequencies, which do appear to be substantially determined by *meaning* (Piantadosi 2014). In storytelling, the word 'love' is much more common than 'hate' because the former is (thankfully) a more common and important topic in our lives. Words with similar distributions, reflecting their occurrence in similar linguistic environments, share similar meanings (see Distributional Semantics, Sect. 1.4.2). Word frequencies also reflect certain kinds of efficiency constraints in cognition. When utterances are gathered together in a large corpus, we find only a few high-frequency words (which are mostly function words like 'the', 'of', 'and', etc.) that make up most of the tokens in the corpus, and lots of low-frequency words, with the frequency distribution bearing a long tail of many words that are seldom used. This reflects *Zipf's Law*, a power law which states that the frequency of a word is inversely proportional to its rank in the corpus. In other words, a small number of words appear very frequently, while the majority are rarely used.

In English, 'the' is of course the most frequent word (rank 1) covering 6.6% of all word occurrences in the 100-million word British National Corpus (BNC). Guess which word comes in at second place? It's 'of' at 3.6%; occurring half as often as the top ranker, with 'and' coming in a weak third, approximately a third as often as 'the', at 2.8%. 'You' is the 13th-ranked word covering 0.8%, and 'not' comes in at rank 20 (at 0.5%), with the percentages then dropping off rapidly.

Function words (like 'the', 'and', and 'not') are the essential building blocks of sentences, and re-using the small set (up to 300 or so in English) of relatively short words repeatedly appears to be an efficiency hack in the language system, where longer and more specialized and interesting words are used, relatively speaking, less often.

> **Discussion of Zipf's Law**
>
> Zipf's Law is a *power law*, where one variable varies as the power of another. It means that the *nth*-ranked word in a frequency table has a frequency proportional to

$\frac{1}{n}$. The law holds for a variety of phenomena, including social networks (where only a small number of nodes are very highly connected), and the distribution of wealth under the extremes of capitalism (where a tiny minority own most of the wealth while the majority own very little). There has been speculation that the law might reflect a fundamental property of self-organizing systems. It is also likely that such a system has a feedback mechanism where common words become more common because they are frequently used.

1.3.2 The Original Language Model

Shannon views the information from the source as being generated by a stochastic process, where each symbol is generated according to its probability of occurrence (in the language at large), and the probabilities of occurrences of previous symbols in the sequence. In particular, he discusses Markoff (or Markov) processes, where each successive symbol in the sequence is generated based on **n-grams**, or word sequences of length n. Sequences of length $n = 2$ are called **bigrams**. As an example, the four bigrams in the sentence "I do love dark chocolate" are "I do", "do love", "love dark", and "dark chocolate". The reader can check how many trigrams there are in that sentence.

Shannon gives examples of NL generated from language models using letter n-grams. Though computed by hand from probability tables, his paper is the first true ancestor, from 1948, of ChatGPT.[4]

The probability of a word appearing after another can be estimated by counting how often the pair appear together and dividing by how often the first word appears overall. If "French fries" appears 50 times in a dataset and the word 'French' appears 100 times in total, then 'fries' follows 'French' about half the time.

This method helps predict likely word sequences, but since some word pairs might never appear in the data. (In fact, a corpus of pre-19th Century writing will not contain "French fries", nor even the modern British equivalent 'chips' used in this sense.) Accordingly, techniques called smoothing adjust counts to avoid assigning to zero the probability of rare combinations. This approach for generating with bigrams, or n-grams for that matter, is an ancestor of current AI methods such as autocompletion while a user is typing as well as for a chatbot predicting what to say next.

---

[4] Shannon's mathematical theory, drawing inspiration from physics, is fundamental not only for natural language processing, including statistical machine translation, but also for computer game playing, circuit design, and cryptography, among other fields.

1.3.2.1 Generating with N-Grams*

To generate with word bigrams, we can use the following formula. The probability P of the next word w_i being generated, given the previous word w_{i-1} is:

$$P(w_i|w_{i-1}) = \frac{\text{freq}(\langle w_{i-1}\ w_i \rangle)}{\text{freq}(w_{i-1})} \quad (1.1)$$

> **Remarks on Eq. 1.1**
> N-gram generation means emitting the most probable first word, then the most probable next word given the previous n−1 emitted words. To illustrate Eq. 1.1, the probability of generating 'fries' (w_i) given the previous word 'French' (w_{i-1}) can be measured by counting how often 'French' is followed by 'fries' in the dataset and dividing that count by the count of how often 'French' occurs. In practice, many n-grams, e.g., "the in", will have a count of zero, and so *smoothing* methods are used to provide non-zero counts. These smoothing techniques include backing off to a lower-order n-gram, or using weighted sums of n-grams of different orders. Unigrams (n-grams where n is 1) too require smoothing, e.g., by adding to the counts, due to zero counts and the long Zipfian low-frequency tail.

1.3.3 Entropy

We now discuss a key concept in Shannon's Information Theory.

Entropy measures how unpredictable or surprising something is. Think of the referee flipping a coin to decide whether Roger Federer gets to serve in a Wimbledon final. In a fair coin toss, heads and tails are equally likely, so a result of heads is only 50% certain. Recall the maxim we described earlier: *n bits corresponds to 2^n choices*. Accordingly, for this coin toss event, which has two choices of outcome, the entropy of the event is 1 bit (since n is 1 in 2^n).

In information theory, the more uncertainty there is about an event, the more the information content. If an event is completely predictable, like the rising of the sun, there is no uncertainty, so it conveys, in information theoretic terms, no information content, and the entropy is zero. This seems counter-intuitive when we think of the meaning of "The sun will rise tomorrow", but bear in mind what we said earlier about the overt and covert presence and absence of meaning respectively.

Let us turn to natural language. Entropy is important for language modeling because it helps us understand how unpredictable a text is, and how well we can compress it. For example, if a text has an entropy of 2.4 bits per character, then a 1,000-character-long text would need at least 2,400 bits (or 300 bytes) to be stored efficiently. The lower the entropy, the more predictable the text, meaning it can be compressed more easily. If a text is full of function words like 'the' or 'and', it has lower entropy because those words appear frequently. But if a text contains unusual or *unexpected* words, it has higher entropy, meaning it carries more information per word.

As an example, what's the next word (say, a noun) one would expect after seeing 'poisoned'? One might think 'apple' or even 'chalice'. While reading China Mieville's novel *Embassytown*, however, one would be surprised to find the rare word 'imago' coming next, not to mention subsequent words in this high-entropy sentence: "The poisoned imago has everyone speaking in cathected symbology." Possessing even higher entropy are the fluid, dreamlike sentences in so-called stream-of-consciousness literature, the extreme case being the often baffling *Finnegan's Wake* of James Joyce.

The flip side of the coin is *redundancy*: if a text has low entropy, it is highly redundant, reducing the cognitive load in processing it. The redundancy of English may be, in some calculations, close to 45%! Naturally, language modeling makes good use of such redundancy when it comes to autocompletion or for chatbots to complete what to say next after a user prompt or after what it has said previously.

1.3.3.1 Calculating Entropy*

Let us start with the **information content** of an event x with a probability $p(x)$, shown in Eq. 1.2:

$$h(x) = -\log p(x) \qquad (1.2)$$

> **Details of Eq. 1.2**
>
> In Eq. 1.2, information content $h(x)$ is measured in bits, with the number of bits of information encoding how 'uncertain', 'unexpected', or 'surprising' the information content is. The minus sign is used since probabilities lie between 0 and 1, and logs of numbers in that range are negative numbers.

As an example of the event x, consider the probability of getting a head on a single coin toss. If the probability of the coin toss $p(x)$ were 0.5, as we would expect, then $h(x)$ would be 1 bit. However, if the coin wasn't fair, but biased to have a lower probability (like the U.S. penny, which is known to have a preference for tails, rising to as much as 0.8 when spun on its side), the number of bits of information would increase.

Let the coin toss be repeated twice. We can compute the probability of getting zero heads, one head, or two heads. Let X be a discrete random variable that stands for 'number of heads' and let $P(X = 0)$, $P(X = 1)$, and $P(X = 2)$ represent, respectively, the probability of 0, 1, and 2 heads. $P(X = 0)$ corresponds to one case TT out of four cases (TT, TH, HT, and HH). Thus, $P(X = 0) = 0.25$, $P(X = 1) = 0.5$, and $P(X = 2) = 0.25$.

1.3 Information Theory and Language Modeling

These are the only alternative possibilities for a double coin toss, and thus the probabilities add to 1. The values of P thus form a probability distribution for the discrete random variable X.

We can now define **entropy** for a discrete random variable X that has n possible values $x_1 \ldots x_n$, in terms of the information content h, shown in Eq. 1.3:

$$\begin{aligned} H(X) &= \sum_{i=1}^{n} P(x_i) \; h(x_i) \\ &= -\sum_{i=1}^{n} P(x_i) \; \log \; P(x_i) \end{aligned} \quad (1.3)$$

Explanation of Eq. 1.3

Equation 1.3 defines *entropy*, a fundamental measure of information. A discrete random variable (call it X) represents an experiment involving multiple (say n) events, each with an outcome that is measured by a real number from 0 to 1. These n events are by definition countable, and therefore the random variable is said to be *discrete*. (A discrete random variable is thus different from a *continuous* random variable like height or temperature, which takes on a potentially infinite number of possible outcomes within a real interval.) The entropy of X is expressed as a sum of products, each of which the product of the probability of the event times its information content (the latter is defined in Eq. 1.2). Like information content, entropy is measured in bits.

In this case of X being the number of heads in the double coin toss, the entropy can easily be calculated by the reader as follows:

$$H(X) = -[0.25 \times (-2) + 0.5 \times (-1) + 0.25 \times (-2)]$$

So the entropy H of the probability distribution of the number of heads in a double coin toss is 1.5 bits, which represents the average amount of information or uncertainty associated with the outcomes. The more outcomes there are and the more evenly distributed their probabilities, the higher the entropy, indicating more uncertainty or variability in the possible results. H is largest when one is completely free in the choice, and reduces to zero when one's freedom of choice is, sadly, gone.

Turning to natural language, the entropy of a text sequence of n tokens $\langle w_1, \ldots, w_n \rangle$ depends on the kind of tokens used, i.e., words, subwords, or individual characters. Let's now consider a discrete random variable that ranges over all sequences of words

$\langle w_1, \ldots, w_n \rangle$ of length n in the language $\mathcal{L}$. Its entropy is defined in Eq. 1.4:

$$H(\langle w_1 \ldots w_n \rangle) = \\ - \sum_{\langle w_1 \ldots w_n \rangle \in \mathcal{L}} P(\langle w_1 \ldots w_n \rangle) \, \log \, P(\langle w_1 \ldots w_n \rangle) \qquad (1.4)$$

Explanation of Eq. 1.4

Equation 1.4 is different from Eq. 1.3 in that the random variable X now ranges over all sequences of words $\langle w_1, \ldots, w_n \rangle$ of length n in the language $\mathcal{L}$. (The underlying view of language here is of course as a stochastic process that generates symbols, governed by probabilities.) If a word sequence is highly probable, its contribution to the overall entropy of a text—or narrative—is low. Too many 'the's in a narrative will lighten the overall entropy. Conversely, rare or unexpected words (including 'imago' and 'cathected') from the lower part of the Zipf distribution help raise the entropy of the narrative, making it more unpredictable or surprising at each step of the generative process.

1.3.3.2 Surprisal

Speaking of cathected symbology, an information theory concept that encapsulates it well is **surprisal**. In the context of language modeling, it measures how unpredictable or surprising a word is given its predecessors in the text.

Surprisal: Technical Details*

The surprisal S of the current word w_t is given by the negative log of the conditional probability of the current word given the preceding words as context, shown in Eq. 1.5:

$$S(w_t) = - \log \, P(w_t | \langle w_1 \ldots w_{t-1} \rangle) \qquad (1.5)$$

1.3.4 Cross-Entropy

Now that we have defined entropy, we can consider **cross-entropy**. Suppose you built a language model that predicts the next word in a sentence. If your model predicts 'chocolate' after "I love …", but people more often follow "I love …" with 'you', your model is erroneous. Cross-entropy helps measure how far off a model is. The more mistakes your model makes, the higher the cross-entropy, meaning the model's predictions need more extra bits to encode the real-world data. If your model perfectly predicts the real-world data,

1.3 Information Theory and Language Modeling

cross-entropy equals entropy. If it's off, cross-entropy is higher, meaning your model needs extra bits to accurately model the statistical regularities in the real world.

In practical terms, researchers have used cross-entropy to estimate the upper bound on the entropy of English. Using a trigram language model trained on the nearly 6 million-word Brown Corpus, the approach of Brown et al. (1992), used the cross-entropy of that model to compute the upper bound on entropy of written English (based on 95 printable ASCII characters), finding it to be 1.75 bits per character.

Further insight on discovering entropy for a language

What are the implications for this discovery of the upper bound on entropy of written English being 1.75 bits per character, or BPC? Recall that n bits corresponds to 2^n choices. So this means that on the average, while typing this sentence, there are no more than $2^{1.75}$ or approximately 3.36 choices to make to predict the next character. Bear in mind that this is a gross average. If we just used the 26-letter English alphabet and digits 0 to 9, the BPC would be even lower, making written English even more predictable.

The entropies of written languages are quite varied. Take Chinese, whose entropy is estimated at no more than 1.3 BPC. Each Chinese character conveys a lot of information, but far fewer characters are needed to convey meanings. Character sequences in natural languages are nevertheless easier to predict than chess moves, which are estimated at up to 6 bits (or 64 choices) per move!

If we turn to bits per word (BPW), the entropies get higher, as there are of course many more words in the vocabulary than there are letters in the alphabet or ASCII. The entropy of written English at the word-level is estimated at 8–10 BPW, which means that up to 1024 decisions may have to be made to predict the next word, on the average, without taking context into account.

Modern language models like GPT-3, however, have written English entropies of around 3–5 BPW, reducing the number of decisions to be made in prediction, in a gross average, from 1024 to (and the next word is:) thirty-two! Language models are able to really exploit context and mining of relationships in training data to arrive at accurate prediction of the next word.

1.3.4.1 Computing Cross-Entropy*

As mentioned in the previous section, we often need to compare a given distribution, or one that has been computed, e.g., by an LLM model, to the actual or true distribution. Cross-entropy measures the number of extra bits needed to represent the target distribution P if the given distribution Q is used as an approximation, shown in Eq. 1.6:

$$H(P, Q) =$$
$$- \sum_{\langle w_1 \ldots w_n \rangle \in \mathcal{L}} P(\langle w_1 \ldots w_n \rangle) \, \log \, Q(\langle w_1 \ldots w_n \rangle) \tag{1.6}$$

> **Explanation of Eq. 1.6**
> Equation 1.6 considers the probability distributions of two discrete random variables P and Q which have the same underlying set of events $\mathcal{L}$. Here $\mathcal{L}$ is the language treated as a set of word sequences, and thus each member of the set is represented by the word sequence $\langle w_1 \ldots w_n \rangle$. This equation is especially of interest when Q is the learned distribution from a language model that approximates P, the true distribution of the language $\mathcal{L}$. (Thus the right-hand-side of the equation samples the word sequences from the true distribution and sums the log of the probabilities from the language model's distribution.) This comparison of the model with the true distribution will also be useful in providing an accuracy measure for training and evaluating a language model, as discussed more below and in Chap. 2.

Cross-entropy can also be defined in terms of Kullback-Leibler (KL) Divergence, shown in Eq. 1.7:
$$H(P, Q) = H(P) + KL(P \parallel Q)$$
$$\text{where } KL(P \parallel Q) = \sum_x P(x) \, \log \frac{P(x)}{Q(x)} \tag{1.7}$$

One nice feature of cross-entropy is that it provides an upper bound on the entropy, so $H(P) \leq H(P, Q)$. Cross-entropy can therefore be used to compute the upper bound for the entropy of an entire language, using a language model for the approximation Q. This was done in the above-mentioned approach of Brown et al. (1992).

1.3.5 Probability of a Sentence

Consider this sentence; what is its probability? Ask a language model, and it will examine each word, 'Consider', 'this', and 'sentence', asking what the chance is of that word (or token) given all the words (or tokens), if any, before it. Multiplying all those chances together gives an overall probability for the entire sentence.

Of course, such an approach has to take prior context into account. That context not only includes the prior words in the sentence, like 'Consider' before 'this', and "Consider this" before 'sentence', but also the prior context before 'Consider', which in our case may be empty or unknown. In general, keeping track of context, based on what has been said or even implied, becomes tricky when the text or conversation becomes long. Restricting the

1.3 Information Theory and Language Modeling

context to be shorter is likely to lose meaning, whereas extending the context can cause processing overload by the language model. Natural language is non-stationary, meaning that the chance of a particular word coming next can depend on context that is not just a few words back but even earlier utterances or paragraphs. After all, a speaker might return to an old topic or a writer may return to an earlier plot thread, reflecting *long-distance narrative dependencies*. We discuss these more in Chaps. 2 and 6.

Ensuring the model can handle these distant connections is a fundamental challenge for LLMs, which nevertheless go very far beyond n-grams to track and attend to longer-range context.

1.3.5.1 Computing the Probability of a Sentence*

The **probability of a sentence** of length L is computed using the chain rule for the probability of all the words (or other tokens) in the sentence occurring jointly together, shown in Eq. 1.8:

$$\begin{aligned} P(\langle w_1 \ldots w_L \rangle) &= P(w_1) \times P(w_2|w_1) \times P(w_3|\langle w_1\ w_2 \rangle) \\ &\quad \ldots \times P(w_L|\langle w_1 \ldots w_{L-1}\rangle) \\ &= \prod_{i=1}^{L} P(w_i|\langle w1 \ldots w_{i-1}\rangle) \end{aligned} \quad (1.8)$$

> **Explanation of Eq. 1.8**
>
> Equation 1.8 defines the probability of a sentence in terms of the probability of the first word occurring, times the conditional probability of the second word occurring given the first, times the conditional probability of the third given the first and the second, and so forth—with the last multiplicand being the conditional probability of the last word w_L given all the previous words in sequence. Clearly, for very long sequences, the context will need to be restricted in some way, without falling back to the simplicity of n-gram models. This is in fact a substantial challenge for language models.

Equation 1.8 uses multiplication for expressing sentence probabilities, but for efficiency, LLMs usually avail of addition instead, by the easy trick of converting to logs, as shown in Eq. 1.9:

$$\log P(\langle w_1 \ldots w_L \rangle) = \sum_{i=1}^{L} \log P(w_i|\langle w1 \ldots w_{i-1}\rangle) \quad (1.9)$$

Abstract View of Language Model*
A **language model** computes the conditional probability of the next token conditioned on context, in this case the sequence of previous tokens generated in the sentence or sequence. We are thus considering an *autoregressive* model, that conditions its next output token on the context of its own previous output. Let's represent the input context (such as a prompt) as the vector X. Let $\langle y_1 \ldots y_n \rangle$ be the sequence of previous tokens generated as output, given the input X. The prediction of y_{n+1}, the next word in the sequence, is determined by the previous words generated, the input context, as well as by θ, the function learned by the model. Here θ will contain all the parameters of the model, which we will learn about in the next chapter. Thus, the LLM computes the following probability for the next token to be generated, shown in Eq. 1.10:

$$P(y_{n+1}|\langle y_1 \ldots y_n \rangle, X, \theta) \tag{1.10}$$

1.3.6 LLM Overview

While n-gram language models are rough approximations useful in certain tasks, what we need are much longer and richer models of context. This is obvious when one plays with n-gram generators, which can very easily generate gibberish. However, two important observations can be made here:

1. As the n-grams are lengthened, the text becomes increasingly fluent, the fluency leveling off at a small n. Fluency involves grammaticality, but does not necessarily mean coherence or meaningfulness.
2. The n-gram output often mirrors the language of the training data—in all its idiosyncrasies of style, register, opinion, etc.

These two properties, of **fluency** and **mirroring**, must not be mistaken for understanding of what is being generated, let alone its truth, implications, etc. Those capabilities need to be assessed separately.

Of course, in comparison to n-gram language models, LLMs are far more sophisticated and powerful. Yet they also inherit these two properties, and similar caveats about human inferences from machine output should be kept in mind.

What is an LLM? A large language model is large because of the vastness of both the training data and the huge number of parameters (internal representations in the form of weights) that these models are able to learn (now approaching trillions!). There are three (open) secrets to the tremendous success of LLMs:

1. The modeling of context in terms of word meanings called **word embeddings** that capture meaningful associations and relationships among words, instantiating a linguistic theory of meaning called **Distributional Semantics**, discussed in Sect. 1.4.

1.3 Information Theory and Language Modeling

2. The dynamic modeling of context using models of **attention** that have learned to focus on what is important. These not only take Distributional Semantics further but are coupled to advanced neural net architectures, most recently the architecture knowns as **Transformers**. This modeling goes very far beyond n-grams to use all of the previous context (and succeeding context, for, say, filling in the blanks). At this level, semantic and syntactic relationships among words (or tokens), as well as hierarchical structure emerges. When combined with information from word embeddings, this results in a deeper understanding of meaning than has so far been available to systems, and that in some cases can rival or exceed, as we shall see in the course of this book, that of a human.
3. Training on such a vast scale, allowing for refinement of the relationships over many iterations, without requiring any human annotation. In other words, the learning is self-supervised and entirely automatic.

Let's consider the sentence in Example 1.1:

Example 1.1 *The android died and her lover died of grief.*

According to our definitions, Example 1.1 is a narrative. Let us see what GPT-4o makes of a part of it (the italics are mine, not GPT's):

Example 1.2

> **User Prompt**: The android died and soon after, her lover . . .
> **Machine**: The android died and soon after, her lover *died of heartbreak.*
> **User Prompt**: How about another, longer ending for that sentence?
> **Machine**: The android died and soon after, her lover *began to dismantle the pieces of her memory, unsure whether to mourn the machine or the dream she had become.*

These completions suggest an emergent understanding of story based on character goals, event outcomes, and causality, a point we will explore much more in this book. Next-token prediction is of interest for many different applications, for example, in computing the fluency of a translation or of a sequence of word guesses from speech recognition, or of a task of filling in the blank when a random word is deleted from the sentence, as in a Cloze test (Taylor 1953). Such prediction is of course especially relevant here in trying to generate a story or complete a text given some previous context of generated text, or a prompt.

As we shall see, the LLM is trained on a vast amount of text data, where the next word is already available in the training data. During training, its own guess for the next word is compared against the actual next word (thus eliminating the need for human supervision) by measuring the cross-entropy loss. An LLM model for next-word prediction has learned, based on the learned function θ, a probability distribution for the possible words that could follow, and the most likely word from that distribution is picked as the answer.

1.3.6.1 Perplexity*

The entropy of a LLM is usually measured in terms of bits per character. Sometimes, **perplexity** is used. It is a measure of how well the model predicts the target text. The perplexity PPL of a LLM is related to cross-entropy in a simple way, providing a useful way of assessing the accuracy of a model, shown in Eq. 1.11:

$$PPL(P, Q) = 2^{H(P,Q)} \qquad (1.11)$$

Perplexity can be computed as the inverse of the geometric mean of the probability distribution over all possible outputs for a given input. If, after generating "Go to", an LLM guesses the next word is 'heaven' with a probability of 0.3 and 'hell' (!) with a probability of 0.7, the geometric mean is the square root of 0.3×0.7, with the inverse yielding a perplexity of 2.18 bits per word. In general, low perplexity is good for a model: the fewer the number of choices, the lower its perplexity, and thus the less confused the LLM is.

1.4 Meaning

Having finished with *all* the technical and/or mathematical details for this chapter, let's now return to the fundamental aspect of NL we mentioned earlier and that forms a central preoccupation of linguistics and philosophy.

1.4.1 What is Meaning?

1.4.1.1 Denotational Semantics

In modern linguistics and philosophy, denotational theories of meaning have viewed words as denoting things in the world, whether they are concrete objects or more abstract ones like numbers. The meaning of a sentence, in this view, is the set of conditions under which the sentence is true. The way the meaning of a sentence is built up from the meanings of words and phrases within it is based on **composition**. In other words, the whole is the sum of the parts, and each sum is built up by systematic rules, like the constructs of formal languages (including programming languages and logics, where the notions of syntax, and to some extent, semantics, are borrowed from linguistics).

In denotational semantics, a crucial distinction is made between sense and reference., following Gottlob Frege—the co-inventor of first-order logic (FOL), or predicate calculus, with Charles S. Peirce. The definite noun phrases "the morning star' and "the evening star" both refer to the same object (called 'Venus', named after the Roman goddess of that name), but each has a different sense. Here the notion of the sense of "the evening star" is, roughly speaking, the idea of a star that appears in the evening.

These notions have been discussed extensively in analytic philosophy and formal semantics, and extended to higher-order logics like modal logic to take into account phenomena involving tenses (such as past and future), modals (like 'will', 'may', 'ought to', etc., which express notions of necessity, possibility, obligation, etc.), beliefs, counterfactuals ("if I hadn't done that..."), etc. In possible-world semantics, which is informed by modal logic, a possible world is any situation that is or could be. The meaning of a sentence, called a proposition, is the set of possible worlds in which it is true.

We will return to possible worlds in Chap. 3, dealing in particular with fictional objects that dominate the sorts of narratives we consider in this book. For now, it is important to note that reasoning in FOL is hard.[5] This places some limits on using such logics to reason at the scale and efficiency required for today's systems. Higher-order logics are even harder to exploit efficiently. However, a simpler logic called propositional logic, as we will see, is something that nets can directly implement.

1.4.1.2 Word Senses

When we consider natural language as it is actually used, ambiguity and imprecision seem ubiquitous. In particular, a given word form often carries many different senses or shades of meaning. Some of these different shades are purely accidental. Consider the dictionary senses of 'bear', from the OED:

1. The animal.
2. To carry; to support or hold up; to produce, yield, give birth to; to push, thrust, press.

That these two distinct senses, called homonyms (which correspond in this particular case to part-of-speech differences) happen to be spelled by the same form, is purely an accident, as they originate historically from different forms. Each of these senses has multiple related meanings specified at considerable length in the OED, called *polysemies*. For example, the first sense can have a related meaning of the flesh of the bear (here the same form is used, unlike the case of 'beef' for the flesh of the cow). The second sense can have many polysemies; for example, to carry can mean to display a visible mark, to conduct oneself in a particular manner, etc. A sense can also have metaphorical extensions, as in a 'bear market' for the animal sense.

Different words co-occurring with 'bear' can bring up other related senses, as in "bearing straight ahead", "bearing children", "bearing in mind", "bearing arms", etc. Some of these co-occurring words are relatively frozen as phrases or idioms, meaning they are treated like single words, and others just are predictable sequences, called collocations.

---

[5] The algorithmic time complexity of FOL reasoning is in the worst case intractable, since it is at least (using a method of proof called resolution) exponential and in general worse than exponential.

In addition to explicit senses we have listed here, there are slang uses, differences of meaning in different stylistic registers, and polysemies involving implicit properties (e.g.,"the spoon in the cup" means a different relation of 'in' from "the coffee in the cup"), etc. There are also subtle cultural or other connotations (think of the connotations of a term like 'nunnery' in Hamlet's "Get thee to a nunnery", or of a name like 'Wall Street', or a date like '9/11'),

Listing these different senses is a significant challenge for a lexicographer involved in constructing dictionaries. The main criteria for allowing a word and distinguishing its senses in a dictionary are frequency of use, context of use (including citations), and word origins. It is not possible map all these different senses over an entire vocabulary for a language to meanings in terms of pre-specified concepts, whether these concepts are symbols in a denotational semantics framework, or entities and relations from an ontology or knowledge graph; at best, it can be attempted for narrow domains. Things get even worse when we consider meanings across languages, as some senses may be differentiated in one language, and not in another, and often the senses seem not to line up across languages.[6]

As noted earlier, the efficiency of NL does come at the cost of ambiguity, but the context often resolves this. In the case of narrative, ambiguity may be deliberate, but the context often helps to clarify meanings. Fortunately for NLP, there is a more practical linguistic theory at hand, that suggests how context can be exploited to arrive at meaning. It is this theory, when coupled with clever machine learning approaches that rely on context from training on giant data sets, that provides much of the power of LLMs in understanding NL meaning. We now turn to this approach.

1.4.2 Distributional Semantics

An important development in modern NLP has been its use of distributional semantics. It is often summed up in the words of Firth (1962): "You shall know a word by the company it keeps". The theory actually derives more from the work of Harris (1970) (see also Sahlgren 2008 who argued that words with similar distributions, i.e., those that appear in similar linguistic environments, share similar meanings. Meaning, Harris believed, had nothing to do with reference; what matters are differences of meaning, which are signaled by differences in distributions. The idea in turn derives from the French structuralist Ferdinand de Saussure, who distinguished between syntagmatic relations, having to do with how words combine in linear sequences, and paradigmatic relations, which relate entities that do not co-occur but can be substituted for each other. Consider the partial Sentence Example 1.3:

---

[6] These cross-linguistic issues have motivated numerous linguists to ask whether meaning is language- or culture-specific, but a discussion of this non-trivial topic is outside the scope of this book.

1.4 Meaning

Example 1.3 *A hunter to the core, he crouched low, aimed and ...*

Here, 'aimed' and the possible filler 'fired' are syntagmatically related because they often co-occur in action sequences. The filler 'fired' is paradigmatically related to 'shot' or 'pulled', as these could replace it as possible choices. Distributional semantics captures both these sorts of relations. Each of these words in common usage will have relationships, reflecting, say, that 'hunter' and 'man' and 'prey' are closely tied together, as are 'hunter' and 'gun', etc. Of course, the words 'hunter', 'aim', and 'fire' are related semantically in sentences at large. 'Hunter', 'crouch', and 'aim' will also be related syntactically based on (subject-verb) relationships discovered from sentences at large. 'He' and 'hunter' are also likely to be related in language use.

In distributional semantics, meaning is derived from context. The different kinds of ambiguity, including homonymy and the kinds of polysemies mentioned above, as well as synonyms, hypernyms (e.g., 'mammal' is a hypernym of 'giraffe'), and collocations, fall under the same mathematical framework. In modern NLP, words are not mapped to pre-specified concepts, but are represented by arrays of numbers called **embeddings**, and words that are related in meaning will be closer together in the resulting vector space. A distributional model captures these relationships, and thus allowing one to infer that 'aim' occurs frequently in similar contexts with 'shoot', 'fire', and 'target', leading to higher probabilities for those words when generating text. The presence of 'hunter' and 'aimed', one would expect, will help disambiguate the next word, slanting the prediction towards 'fired' or 'shot'. Note that although the relations are grouped under Distributional Semantics, they may very likely contain syntactic, morphological, or other relations in the mix. For example, subject-verb dependencies ("the dog eats" versus "the dogs eat") will likely be found among the relationships, as we will see in Chap. 2.

1.4.3 Do Machines Really Understand?

Given that we have discussed some aspects of meaning, it is worthwhile asking to what extent a machine understands our input or what it has emitted. This question applies not only to LLMs, of course, but to any NLP system, including one for machine translation; the more general question of whether machines can think has been a subject of lively philosophical discussion for centuries. Delving into the rather fascinating history of this topic is outside the scope of this book; I will confine myself to brief remarks here.

The commonsense answer to whether machines can understand is that any ascription of understanding to a machine is due entirely to anthropomorphism on our part. The philosopher Searle (1990) has made a similar point in his *Chinese Room* argument, where he points out that a person who doesn't know a word of a foreign language could consult a rule-book in a language he does understand to perform correct translations that are indistinguishable from those of a native speaker of the foreign language. If the person were replaced by a computer, the latter would pass the Turing Test (Turing 1950) for whether a machine can be said to think (namely, when the machine's responses in question-answering are indistinguishable's from a human's)—but without having understood even a word of the foreign language! Thus, one might conclude that a machine really doesn't understand.

However, leaving aside the fact that the Turing Test can be too easy to pass, Searle's argument has been rebutted (I think effectively) by Pinker (1997). The latter points out that Searle's argument seems similar to traditional arguments against machines being able to 'fly', revealing our commonsense reluctance to use a word when the situation deviates from stereotypical conditions. At the risk of sounding glib, Pinker argues that 'understanding' is largely a matter of 'semantics'.

Philosophical discussion is however not that useful for system developers or those who seek to understand the true capabilities of Gen AI systems. Benchmarks are far more useful. Developing the right sort of large-scale benchmark tasks to test LLMs, which would require a great deal of linguistic and well as commonsense inference, isn't easy. Some tasks, like standardized tests at which machines are starting to get the upper hand over humans, do not pinpoint the types of inferential understanding a narrative researcher would be interested in. More specialized reasoning tasks, including the useful Story Cloze test (Mostafazadeh and Nathanael 2016), currently only touch the surface of narrative understanding and generation capabilities. In exploring narratological capabilities of machines compared to humans, my experiments, informal though they are, can hopefully inform the next generation of benchmarks.

1.5 Deep Models and Human Linguistic Abilities

We now turn to comparisons of Deep Models used in Generative AI and the language processing capabilities of humans. In today's neural nets, there is no *explicit* representation of morphology, syntax, semantics, or discourse relations. This begs the question of whether the representations posited over the centuries by linguists and cognitive scientists are even valid models. In other words: do these linguistic representations posited for human language have correlates in the human brain?

The answer, without digressing deeply into the psychology and neuroscience literature, is a qualified yes:

- Neuroscience studies (He et al. 2022) have revealed that when humans listen to stories, the similarity between words as measured by brain EEG response seem to correspond to the similarity between word embeddings in neural models. This, the authors conclude, reflects the common influence of the structural properties and statistics of natural language on both neural models and the brain.
- This similarity between nets and humans extends to syntactic structure in language as well. As Manning et al. (2020) discovered, there is a remarkably close correspondence between the neural activations of LLMs doing a Cloze task and patterns involving the

syntactic structures of utterances. Since syntactic structures are implicated in human brain processing of language, this provides indirect evidence of a mapping between LLM activations and human processing. LLM developers such as Wang et al. (2019) have also directly been able to guide the LLMs to focus on such syntactic patterns.
- Researchers such as Tuckute et al. (2024) have compared, when processing a sentence, the internal neural net activations of LLMs with fMRI activations in human brains, and found similarities.

1.5.1 Biological Validity of Neural Models

Neural nets of course have an obvious biological inspiration. In our bodies, neurons integrate multiple incoming signals, both excitatory and inhibitory, before deciding whether to fire an action potential, for which the cell membrane voltage must cross a threshold value. While this sort of information integration is the essence of today's neural computing, the analogy with biological processes more or less stops there. It's worth pausing to clarify this point, to avoid the biological hype that accompanies a lot of articles on neural nets.

A biological neuron's firing has a complex timescale, with different timings for the electrical versus chemical processes involved, which in turn have different energy consumption requirements. The firing also depends on multiple levels of feedback as well as the neuron's current state (the latter is called recurrence). In turn, the firing propagates to specific local areas. Incorporating timings and recurrence alone leads to much more computationally complex models, whose implementation can require specialized hardware, all of which is currently only available in a few research settings.

1.5.2 Deep Learning Versus Human Language Acquisition

The advances of Generative AI have sidestepped the carefully honed biological adaptations that resulted over millennia in human language. In other words, a robot today can benefit from the same evolutionary adaptations without having to go through any biological evolution involving the embodiments that humans actually use to produce and understand speech and language. Instead, it has harvested the indirect products of human evolution in terms of huge datasets of linguistic data. In comparison, the interactive process that guides a child's language acquisition, with far less data, is carried out at a relatively slow pace, so that by age five, a child can recognize about 10,000 words (Shipley and McAfee 2024). This sort of acquisition process is completely side-stepped by the LLM, which relies of course on training on orders of magnitude more data scraped from the web. In terms of language capabilities, the LLMs thus represent a revolutionary and truly astounding development in artificial systems.

1.5.3 Language Universals and Information Theory

Let us now turn to commonalities across human languages, and see whether those commonalities have parallels in neural nets.

Natural language is a recent evolutionary adaptation, dating back only a few hundred thousand years. It may have evolved as a solution to challenges in communication. Unlike animal communication, with which human language shares many commonalities, NL is highly productive, can reference abstract concepts, and has common elements across the many languages that exist or once existed. Studies of human languages across the world by Greenberg (1966) and others have emphasized common properties, some of which have been posited as **language universals** that hold across all or a substantial grouping of languages. For example, languages with dominant Verb-Subject-Object (VSO) order have the adjective after the noun. In languages with prepositions, the genitive almost always follows the governing noun, while in languages with postpositions it almost always precedes the noun. Languages with dominant VSO order are always prepositional. These claims are hardly without exceptions.

The most important language universal is probably the fact that the internal structures of NL display a unique property of **compositionality**, where the information content of a larger unit (e.g., a sentence or word) is a systematic function of the information content of the parts (Howe 2012) (see also Futrell and Michael 2022). Any artificial system which mimics natural language, one would think, should possess this characteristic. However, today's LLMs, which are mainly neural and sub-symbolic, do not use or directly exhibit any such compositionality, which has to be inferred, if at all, as a side effect (Donatelli and Alexander 2023).

The notion of entropy central to LLMs and information theory is also directly relevant to the study of universals across languages. As an example, a language like English follows for the most part a SVO word order, this rigidity being measured in terms of lower entropy (by counting the frequency of that pattern in a corpus divided by the total number of such patterns). Other languages, like Russian or Tamil, have more flexible word orders, corresponding to higher entropy. As Levshina (2021) argues, more rigid word order in a language can contribute to more efficient communication by reducing cognitive loads, while more flexible word order is accompanied by case marking and other grammatical devices to reduce ambiguity. Some languages may prioritize processing efficiency by developing rigid word order, while others might balance flexibility with the use of additional grammatical tools to maintain clarity.

1.5.4 Efficiency in Information Theory and NL

Efficiency is a key consideration in information theory, and it does seem highly relevant to NL. As pointed out by Gibson et al. (2019), it appears to be an important evolutionary criterion

1.5 Deep Models and Human Linguistic Abilities

for structuring natural language. We have already seen how Zipfian word frequencies may reflect cognitive constraints on efficiency.

We saw earlier that *surprisal* is a measure of how unexpected a word or token can be given the prior context. For communicating efficiently in NL, researchers have posited the hypothesis of *Uniform Information Density* (UID). UID holds that speakers will try minimize large jumps in surprisal across an utterance, to keep the information content distributed as smoothly as possible. UID has been shown empirically to operate in both production and comprehension of utterances (Meister 2021). However, this hypothesis may be violated in narratives to create not just creative stylistic innovations like we saw with China Mieville's *Embassytown*, but also unexpected plot twists and other surprising developments.

Efficiency in NL does come at the cost of *ambiguity* and *imprecision*. Nevertheless, this cost is largely offset in practice. Though NL may be ambiguous in textbook examples, in practice the context often disambiguates it, a principle that underlies much work in LLMs. As for imprecision, this allows for qualitative and vague descriptions of, for example, spatial and temporal positions and relationships through prepositional phrases, verbs expressing motion, and other constructs (Mani and James 2012). (Of course, when the need arises, NL can use technical language to make things precise.) Evolutionary pressures may well have contributed to both compression of linguistic forms to ensure shorter, more easily articulated utterances, as illustrated in a wide variety of phonological phenomena involving sandhi and assimilation. As for the ubiquitous redundancy in NL, this may have evolved to make sure messages can be transmitted correctly in the presence of noise.

Since our everyday experience of life involves temporally ordered events, related to goals and outcomes that have emotional value, stories too must reflect this experience. It is therefore a reasonable hypothesis that narratives that parallel real-life experience will be processed and remembered more efficiently. Indeed this hypothesis has been tested in various ways, see Graesser et al. (2002). Some related experimental results are discussed in Chap. 5. For the time being, it's worth noting that narrative structures as well as conventions seem to incorporate many of these fidelity-to-life properties. Too much breaking of conventions and the narrative becomes harder to process, judged incoherent or uninteresting.

However, while efficiency seems a relevant characteristic of human language and narrative, many forms of narrative, including the earliest oral poetry, can be deliberately verbose. Entertainment is clearly an additional function of narrative, which goes beyond the goals of the Shannon model of communication, though Warren Weaver argues in Shannon and Weaver (1949), without much elaboration, that success in communication across a channel also includes the receiver's response and behavior, which can involve aesthetic responses.

Thus, we are faced with a rather paradoxical situation: as we will discuss in this book, LLMs can understand, produce, and potentially reason about NL narratives based on formulating stories as an optimization problem that gets at the most probable next item to follow given the context. But a narrative itself appears to have properties that resist such a formulation.

1.6 AI Limitations

1.6.1 LLM Limitations

As the reader may already be aware, the current crop of Gen AI technologies have proved useful in many different areas, but they are also prone to numerous problems:

Truthfulness
With LLMs, a particular issue is their lack of any notion of *truth*. We have already seen that information theory lacks any notion of truth, with successful transmission of information being the aim. It therefore is somewhat unfair to suddenly expect truthfulness to be an objective within a neural system. Since LLMs aren't guided in their built-in mechanisms by Asimov's Laws of Robotics (or other such behavioral protocols), they have no direct responsibilities as artificial agents. It is thus left to, for example, chatbot wrapping software, to introduce such laws and responsibilities as guardrails, which have for various reasons, turned out to be both naive and weak.

Bias
This is perhaps the most serious problem of all, which can result in amplifying discrimination, prejudice, misinformation, and many other horrible qualities. This is a weakness not so much of the LLM per se but of the humans who provided the training data and those who allowed the system to be trained on it. Another aspect of LLM bias is neglect and marginalization due to their dependence on large corpora. For many languages and cultures, including those of the past, there are no such corpora, and thus LLMs don't apply to them.

Hallucinations
LLMs also tend to hallucinate, for example, generating fake citations or mistaking their own output for reality. The tendency to hallucinate has been attributed to ingesting bad data, *overfitting* to the training data (i.e., being unable to generalize beyond it), and insufficient understanding of context. There is also a cascading effect: once the LLM has predicted a hallucinated word at a given step, the odds of extending the hallucination at subsequent steps are raised. This is a huge problem, with all sorts of social consequences, but it may eventually admit of a technological solution. Somewhat encouraging results have been obtained through Retrieval Augmented Generation (RAG), see Chap. 2.

Opacity
The representations used by neural nets are numeric and devoid of much symbolic content that can help explain in human terms why the net arrived at a particular decision. If we grant that machines and humans use entirely different representations to understand and compose sentences, we are faced with the problem of machine opacity. (Humans are sense-makers, whereas machines care little for that, aiming for efficient computation.) There is, however,

considerable research on hybrid neurosymbolic systems aimed at addressing this problem, some of which is explored in Chap. 6.

Multiple Failure Modes
Due to an LLM's fluency and mirroring, a human can be easily misled into making incorrect inferences ascribing greater knowledge to the machine than is deserved; the user is then surprised or frustrated when it fails. It can do so in many ways, making the most rudimentary mistakes.[7] For example, GPT fares very poorly compared to spreadsheets like Excel. Mistakes that it makes in elementary arithmetic would shock even the most basic calculator, let alone a schoolchild. However, language modeling was not originally intended for arithmetic, even though the underlying nets carry it out accurately. An LLM can often forget instructions in session memory, and get confused when the session extends too long. This is probably a reflection of its limitations with handling context, and may also be an artifact of the way the LLM is wrapped in, say, the chatbot application. Due to its non-determinism and lack of any notion of truth or logic, the system can also easily contradict itself over successive turns.

Narrative Limitations
From the presentation so far, it seems clear that an LLM can generate a narrative of any length a word at a time. The machine takes into account the vast trove of information in its training data, as well as the previously generated context, including linguistic and other emergent relations between words in that context. This can be effective when it is generating the next sentences of a story based on a previous sentence from anywhere in the story. Human writers aren't limited to that, though. They do not necessarily create their stories from beginning to end; they often know the ending, and sometimes they even write from the middle out. This is discussed more in Chap. 6. One might think that this would preclude the LLM from generating an entire middle based on the beginning and the ending (Riedl 2021). However, this is mitigated by the fact that the LLMs have many stories in their training data, allowing them to develop implicit knowledge of narrative arcs. Nevertheless, LLMS still have serious limitations:

- They do not fare well on generating longer texts, with their outputs becoming incoherent at higher lengths.
- We also saw already that there is nothing in the probabilistic model that addresses longer-distance narrative dependencies. Thus constraints from a fall from grace at the end of a story cannot be used to set up preparatory actions earlier on for the character to experience.
- Due to overfitting, the LLMs may not generalize that well to completely new story patterns.
- LLMs are also limited in terms of input length. To get such a system to read an entire novel at a time, let alone a long short-story, is infeasible.

---

[7] There are many human factors that enter in when one knows one is interacting with a machine, including anthropomorphizing it and being repelled by how eerily human the machine seems (the 'uncanny valley' phenomenon that arises with robot interactions).

1.6.2 Problems not Specific to LLMs

It is worth noting that LLMs aren't by any means the only computational tools for understanding and generating stories. Earlier computational approaches were based on hand-crafted rules, but were restricted to particular domains, and in general proved to be rather brittle. Today's machine-learning approaches, especially those that don't require a human labeling examples for training, are much preferred to those earlier rule-based approaches. There are several non-LLM machine-learning approaches including reinforcement learning algorithms that are highly applicable to narrative problems such as story generation. Our main focus, however, is on LLMs due to their outstanding evaluations on state-of-the-art benchmarks for many tasks involving human language—though, as I have pointed out, they do not include sophisticated narrative benchmarks.

There are nevertheless other problems that aren't specific to LLMs but that pertain to AI as a whole. I will just touch on these as detailed discussion would digress from the main arguments:

Training Costs
Machine learning systems for big data are very expensive given their compute-intensive nature, putting training of large models (which have billions or trillions of parameters) out of reach of communities that aren't that well funded. The details of the computational costs and how to amortize them are interesting in themselves and also the focus of a vast industry, but is outside the scope of this book. However, it is worth mentioning, as Eldan and Li (2023) has shown, that insights and emergent knowledge can also be seen in LLMs trained on smaller datasets.

Energy Costs
Training GPT-3 is said to have, with some caveats, consumed 1287 MWh (megawatt-hours) of electricity. That is guesstimated to be enough to power an American household for more than a century. The infrastructure of the data centers used can also be extremely energy demanding. As it happens, Meta (the parent company of Facebook) recently put out a call for proposals for nuclear reactors to fuel its energy needs. In a planet that is in apparent climate free-fall, having a ravenous AI technology which dramatically ramps up energy demands is not that sustainable a prospect.

Artificial Agency
Yet another set of problems arises because AI systems are serving as artificial agents. The machine has no knowledge of social settings or cultural references outside its training data, and no cultural context outside that. There is no cultural sensitivity, nor even any knowledge of ethics. Nor does the agent have any embodiment, real emotion, or grounding in nature. These are serious liabilities when it comes to grappling with narrative. However, the lack of embodiment may change once wetware computing advances further. Also, what is not genuine about simulating emotions? After all, emulating or imagining other's emotions is

what underlies our own empathy, especially when reading fiction and evaluating characters. We shall have more to say about such evaluations in Chap. 3. Discussing emotion in more detail, unfortunately, is outside the scope of the current book.

Authorship Attribution
Since Gen AI-derived output can be increasingly difficult to distinguish from human output by either humans or machines, there is the increasing likelihood of human authors indulging in plagiarism and dishonesty, whether in the (now all-too-common) case of students passing off GPT texts as their own or in scientific publications submitted for peer review. In the latter case, an author's use of Gen AI can include text polishing, as well as (likely incomplete or hallucinated) literature reviews or summaries, and (possibly inaccurate) scientific analyses, not to mention completely fabricated results, to name just a few possibilities. Almost all serious peer-reviewed publications now insist that the use of AI be disclosed along with paper submission, and have stringent guidelines on the use of AI. (Peer review itself should also be free of the use of AI, so that papers can be assessed honestly and without bias.) The same applies to other creative works, including narrated stories, which we discuss more in Sect. 1.6.4, Society as a whole has to decide how intellectual property is to be managed in these situations.

Safety
AI has already produced violent and offensive content, unethical deep fakes, disinformation, weapons, and other such pernicious evils welcomed by bad actors. Pertinent to this are conspiratorial narrative networks. While this is not the focus of this book, I briefly examine the topic under future research in Chap. 7.

Ethical Guardrails
Ethical frameworks for applying information technology have been in development for a while. For example, the framework of Wright (2011) is derived from biomedical ethics principles. It uses a questionnaire for the technology developer or stakeholder to consider specific ethical issues, including: Autonomy, Dignity, Informed Consent, Safety, Social solidarity, Value-sensitive design, Sustainability, Use limitation, Confidentiality, Anonymity, and Privacy. It is not clear how well it can be implemented in practice. A paper by Ebell et al. (2021), written in response to the silencing of research about ethical issues in a well-known big tech company, discusses the challenges of implementing ethics in corporate-funded research, given the propensity for conflicts of interest and systemic biases. Protective measures they suggest include whistleblower mechanisms, requiring statements in papers of any conflicts of interest, providing a platform to "under-represented, discriminated-against, or censored members" of the community, and having records of when and for what purpose certain algorithms are used.

1.6.3 Social Limitations

Narration has its origin in oral narratives, which have a social function. Today a lot of narrative formation is online, with discussion groups collectively creating narratives, whether meritorious or conspiracy-based. In this narrative era, is it too fanciful to think of families of robots, lacking any organic culture, gathering together telling tales to each other and sharing them with humans? It's easy enough to write a program involving LLM cooperation, inspired by collaborative poetry like the Japanese *renga*, which are chain poems where poets collaboratively extend each other's stanzas. One LLM may generate the first sentence of a story, and another differently-architected LLM, having read the sentence, can add a second sentence, the alternation continuing until the story is done. The LLMs could be trained together, one spontaneously consulting the other when it gets into difficulty. Robots engaging in continuous learning would thus be teaming up to produce narratives. The *CollabStory* (Venkatraman et al. 2025) project has over 32,000 stories generated collaboratively by up to 5 LLMs. Each story segment is generated from a Reddit writing prompt by a single machine author, which "passes the narrative baton to the next, completing the storyline part by part in a sequential manner." The writing prompts address beginnings, middles, and ends. (However, the goal of this project is plagiarism detection, namely to detect if the text is written by a particular LLM, rather than evaluating the quality of the stories.) Home assistance robots might provide this sort of entertainment to humans as well as perhaps each other, if their social sides were more developed. This presupposes that the robots can in fact generate entertaining narratives, display the elements of narrative understanding, and simulate the sorts of reactions that humans have. Can they? The detailed answer is the gist of this book.

1.6.4 Limitations in Creativity

Today, Generative AI has produced paintings exhibited in art galleries and music performed by orchestras, as well as numerous narratives and poems that seem indistinguishable at times from human creations. AI systems are now assisting humans in their creative efforts, including scientific discovery and brainstorming. But are these AI systems creative themselves? Creativity can be characterized in terms of inventiveness, resourcefulness under constraints, violation of conventions, and usefulness and novelty of the results. However, what makes a work creative also seems subjective and culture-specific.

The field of Computational Creativity studies "The philosophy, science and engineering of computational systems which, by taking on particular responsibilities, exhibit behaviors that unbiased observers would deem to be creative." Colton and Wiggins (2013). There are a variety of computational theories of creativity like the tripartite framework of Boden (2003), which involves combinational, exploratory, and transformational creativity, and the model of Wiggins (2006), which views creativity as a structured search through conceptual spaces. To encourage diversity and autonomy, systems have used evolutionary computation including genetic algorithms (which widen the space of possibilities using mutation and crossover, and fitness-based selection), as well as conceptual blending and analogical reasoning, which allow for distinct domains to be merged to create novelty. More control over the output and its coherence is provided by domain-specific rule-based approaches, which allow for aesthetic and other constraints. In recent years, neural approaches including LLMs and Generative Adversarial Networks (GANs) have made great headway, including systems like Google's Magenta, Open AI's MuseNet and DALL-E2, and Stability AI's Stable Diffusion.

However, there is no general conception of what creativity metrics could be use to guide or evaluate a system, nor of what fundamental operations may underly creativity within or across domains. Studies often rely on experiments derived from Turing Tests, or domain-dependent expert judgments. Even more concerning is the fact that much of Generative AI could be viewed as inauthentic, as it borrows from human creations in its training data.[8] Finally, given that these are artificial agents lacking any intrinsic sense of culture or ethics, as mentioned in Sect. 1.6.2, safety and ethics become especially important when AI is allowed to run free.

Despite these caveats, we nevertheless examine some highly creative generated stories in Chap. 6.

1.6.5 Human Limitations

While it is easy to find fault with machines, an even larger problem arises due to information produced by humans that is present in the training data. While scientists are by and large committed to discovering truths about nature, information about science and the world at large, including discussions in formal and informal conversations, has always included unverified and often unverifiable opinion—and in the case of non-science discussion, the presence of mythology and false belief. These are part and parcel of human thinking and culture, and as a result, a key ingredient of fictional as well as non-fictional narrative.

1.7 Summary

This chapter has introduced storytelling, defining concepts like *Narrative* and *Narrative structure*, using insights from the field of Narratology and literary studies. The reader should have grasped the fundamentals of *Information Theory*, including concepts of *entropy* and *cross-entropy*, and the basic techniques that underlie *n-gram language models*. After pointing out the distinguishing properties of *LLMs*, the chapter went on to discuss meaning, including different word senses, providing a basic introduction to logic-based approaches to meaning and *Distributional Semantics* (the latter, as we have noted, is implemented in

---

[8] This raises economic and social concerns, and lawsuits on rights infringement are expanding.

LLMs). After approaching the deeper philosophical question of whether machines can really understand, we examined different strands of research comparing Deep Learning models with human linguistic abilities, addressing biological validity, language acquisition, and language universals. Finally, I outlined several important limitations of LLMs and AI, as well as humans.

1.8 Book Overview

1.8.1 Chapter Previews

This book addresses the extent to which well-studied properties of narrative can be exhibited and understood by machines. Chapter 2 reviews Neural Networks and Transformers, providing a technical overview of LLMs such as found in Chat GPT and its various rivals. A key insight is the crucial role implementations of the linguistic theory of word meaning (in the form of Distributional Semantics) plays in these powerful LLMs. Chapter 3 introduces key concepts from narratology, shedding light on them using LLMs. Chapter 4 discusses plot and its ramifications. Chapter 5 examines a further narratological concept: time and its patterning. While generation is discussed in earlier chapters, Chap. 6 examines story generation algorithms in detail, exploring examples and shortcomings in that respect using LLMs. Chapter 7 summarizes the unique contributions of this book, and examines future prospects, including the application of these techniques to video narratives and movies.

In each chapter, I explore the extent to which LLMs are able to detect and explain plausible and implausible violations of narrative structure. These violations include causal gaps, loose threads, cliffhanger or unsatisfactory endings, incoherent point-of-view shifts, pointless atemporality, implausible characters, lack of motivation for actions. While an LLM can predict the likelihood of alternative next words as well as sequences, detecting and explaining these violations, invoking along the way notions of plausibility, requires tapping into the machine's emergent knowledge of narrative structure. And if it turns out that a machine lacking in emotion and culture can make similar judgments of plausibility as we humans do, it does beg the question as to how intrinsically human these judgments are. This approach is inspired by the Story Cloze test (Mostafazadeh and Nathanael 2016), where the task is to determine which of two crowdsourced micro-story endings is the right ending, which is often used as a benchmark for LLM evaluation. LLMs tend to fare rather well on this task, with an accuracy of 90% that is comparable with humans. However, that latter test is far simpler than some of the questions about literary narrative posed to the LLM in this book.

My assessments don't offer the same satisfaction that a quantitative empirical evaluation would provide, but constructing a large-scale test set and benchmark with such narratological examples is a very substantial task beyond the scope of this book; however, see the narrative corpora I have annotated, mentioned in Sect. 1.8.3. Further, though my results are anecdotal, they have the advantage of providing in-depth insight into individual cases in a way that a large-scale evaluation may not be able to provide.

1.8.2 Protocols for Testing Examples

Throughout this book, I will use examples with my prompts and LLM zero-shot output—meaning that unless otherwise indicated, I provide only one set of instructions with no training examples in them, and with input texts that are most likely unseen. To avoid having the machine learn from previous interactions with me, my zero-shot examples are executed after clearing the session memory for each interaction. Since regenerating the answer or trying the same question in a new session will not provide exactly the same answer, due to both the non-determinism of the algorithms and differences in versions and their training data, one has to decide which answers to include in the book as examples. To avoid cherry-picking, and to make sure the answer is not a fluke, I check if the answer or something very similar can be regenerated several times.[9]

It's possible, however, that some similar passage or part thereof has been seen in the LLM's vast training corpus. One could anonymize the story to guard against this by using various techniques, including replacing proper names by letters, paraphrasing, translation and back-translation. Such methods are undoubtedly useful in testing whether the machine is actually carrying out reasoning or relying on parametric knowledge that has been 'memorized'. This is a larger challenge to try and pin down, and I don't think it's worth the trouble for this effort.

As for which LLM is used for a given example, it's best to report where possible a fixed LLM model for consistency. At the time of writing, GPT-4o seems to provide the best overall results. To see if similar results are within the scope of other LLMs, I also try the same problem with other Foundation LLMs (including different GPT models, Meta-Llama-3.1-70B-Instruct, Google Gemini Advanced, a variety of models available via Hugging Face, and Anthropic's Claude 3.5 Sonnet). If the results seem beyond the other LLMs, I note it in the text. When probing the internals of the networks, if the API's for certain models fail to expose the desired internals, I use in my API calls the maximum strength model that does expose them, and report the model accordingly.

1.8.3 NarrativeML

At the end of each chapter, to help clarify concepts, I extend a data structure to represent the distinctions that have been introduced. This is expressed in the form of an annotation scheme for narrative corpora called NarrativeML (Mani 2014). The scheme is rather like a fleshed-out fabula, with the details of the narrative relations, characters, their goals, events,

---

[9] I have also carried out simple sanity checks such as shuffling the text, swapping character references (e.g., changing 'he' to 'she' in the middle, etc.), and introducing random words or names to see if the plausibility judgments are sensitive to such perturbations. Such sanity checks are easily passed by the LLM.

times, and spatial, temporal relations all specified, along with information about the audience reaction to the narrative.

At the same time, NarrativeML can stand as a representation for different discourses or texts which have the same detailed structures. It is thus less abstract than the story plot. The tags in the annotation scheme are in turn tied to an ontology of abstract concepts pertinent to narrative. The ontology is presented informally, while the annotation scheme is described in terms of an Extensible Markup Language (XML) Document Type Definition (DTD). NarrativeML is a fairly 'lightweight' annotation scheme, and yet it requires a lot of effort to annotate by humans. As I will show, LLMs are now able to carry out this markup 'reasonably' accurately and effectively. The narratological aspects of examples discussed in this book are thus automatically marked up in NarrativeML using a LLM. Chapter 7 discusses some challenges in assessing inter-annotator reliability of NarrativeML annotation. The Appendix describes the annotation guidelines for humans.

A word on storytelling. The majority of the book is focused on LLMs which have to do with both analysis and generation, except for Chap. 6, which is focused on story generation. Generation has its own set of structuring and background knowledge requirements. However, from a computational standpoint, though story analysis and generation may be different in terms of processing components and knowledge requirements, it does not preclude their sharing common structural representations of narrative at particular levels of abstraction, as NarrativeML does.

There are also several fiction corpora that have been machine-annotated in NarrativeML that I have made available online here: https://tinyurl.com/suproc. Included are the 5-sentence Story Cloze (or ROC) stories (Mostafazadeh and Nathanael 2016) studied earlier by the community that are annotated automatically in NarrativeML, as well as NarrativeML for a sample of stories generated automatically from the 300K-story Writing Prompts dataset (Fan et al. 2019) culled from Reddit. These corpora constitute the first time an expressive formalism that captures events, times, characters' goals, actions, and outcomes, along with details of plot, is available on a large scale for narratives. This large resource, though not validated by humans, may be used for studies of narrative by scholars in other disciplines, as well as for further training (via what is called fine-tuning, to be discussed in Chap. 2) of LLMs and other Gen AI tools.

References

Bal, Mieke (1997). *Narratology: Introduction to the Theory of Narrative*. Toronto: University of Toronto Press.
Boden, Margaret (2003). *The Creative Mind: Myths and Mechanisms*. Milton Park, UK: Routledge.
Brown, Peter F., Stephen Della Pietra, Vincent J. Della Pietra, Jennifer C. Lai, and Robert L. Mercer (1992). "An Estimate of an Upper Bound for the Entropy of English". In: *Computational Linguistics* 18, pp. 31–40. URL: https://api.semanticscholar.org/CorpusID:18511291.
Chatman, Seymour (1980). *Story and Discourse: Narrative Structure in Fiction and Film*. Ithaca: Cornell University Press.

References

Colton, Simon and Geraint A. Wiggins (2012). "Computational Creativity: The Final Frontier?" In: *European Conference on Artificial Intelligence*. URL: https://api.semanticscholar.org/CorpusID: 5880786.

Donatelli, Lucia and Alexander Koller (2023). "Compositionality in Computational Linguistics". In: *Annual Review of Linguistics* 9.1, pp. 463–481.

Dubourg, E. and N. Baumard (2022). "Why and How Did Narrative Fictions Evolve? Fictions as Entertainment Technologies". In: *Front. Psychol.* 13, p. 786770.

Ebell, Christoph et al. (2021). "Towards intellectual freedom in an AI Ethics Global Community". In: *AI and Ethics* 1.2, pp. 131–138. URL: https://doi.org/10.1007/s43681-021-00052-5.

Eldan, Ronen and Yuanzhi Li (2023). *TinyStories: How Small Can Language Models Be and Still Speak Coherent English?* Preprint arXiv:2305.07759.

Fan, Angela, Mike Lewis, and Yann Dauphin (2019). "Strategies for structuring story generation". In: *Proceedings of the 57th Annual Meeting of the Association for Computational Linguistics (ACL'2019)*, pp. 2650–2660. URL: https://aclanthology.org/P19-1254/.

Feuerriegel, Stefan, Jochen Hartmann, Christian Janiesch, and Patrick Zschech (2023). "Generative AI". In: *Business and Information Systems Engineering* 66.1, pp. 111–126. URL: http://dx.doi.org/10.1007/s12599-023-00834-7.

Firth, J. R. (1962). *Studies in Linguistic Analysis*. Oxford: Wiley-Blackwell.

Forster, E. M. (1956). *Aspects of the Novel*. New York: Harcourt.

Futrell, Richard and Michael Hahn (2022). "Information Theory as a Bridge Between Language Function and Language Form". In: *Front. Commun.* 7, pp. 657–725. URL: https://www.frontiersin.org/journals/communication/articles/10.3389/fcomm.2022.657725/full.

Genette, Gerard (1980). *Narrative Discourse*. Ithaca: Cornell University Press.

Gibson, Edward, Richard Futrell, Steven T. Piantadosi, Isabelle Dautriche, and Kyle Mahowald (2019). "How efficiency shapes human language". In: *Trends Cogn. Sci.* 23, pp. 389–407.

Graesser, Arthur C., Brent Olde, and Bianca Klettke (2002). "How does the mind construct and represent stories?" In: *Narrative Impact: Social and Cognitive Foundations*. Ed. by Melanie Green, Jeffrey Strange, and Timothy Brock. New Jersey: Lawrence Erlbaum.

Greenberg, Joseph H. (1966). *Universals of Language, 2nd edn*. Cambridge, MA: MIT Press.

Harris, Zellig (1970). *Papers in Structural and Transformational Linguistics*. Dordrecht: D. Reidel.

He, T., M. A. Boudewyn, J. E. Kiat, K. Sagae, and S. J. Luck (2022). "Neural correlates of word representation vectors in natural language processing models". In: *Psychophysiology* 59.3. URL: https://doi.org/10.1111/psyp.13976.

Herman, David (2005). "Quantitative methods in narratology: a corpus-based study of motion events in stories." In: *Narratology beyond Literary Criticism*. Ed. by Jan Christoph Meister. Berlin: De Gruyter.

Howe, Stephen (2012). "A re-examination of Greenberg's universals". In: *Fukuoka University Review of Literature and Humanities* 14.1, pp. 209–253.

Jockers, Matthew and Rosamond Thalken (2020). *Text Analysis with R for Students of Literature*. Berlin: Springer Nature.

Levshina, Natalia (2021). "Cross-Linguistic Trade-Offs and Causal Relationships Between Cues to Grammatical Subject and Object and the Problem of Efficiency-Related Explanations". In: *Front. Psychol.* 12. URL: https://www.frontiersin.org/journals/psychology/articles/10.3389/fpsyg.2021.648200/full.

Mani, Inderjeet (2014). "Computational narratology". In: *Handbook of Narratology*. Ed. by Peter Huhn, Jan Christoph Meister, John Pier, and Wolf Schmid. Berlin: De Gruyter, pp. 84–92.

Mani, Inderjeet (2016). "When Robots Read Books". In: *Aeon*. URL: https://aeon.co/essays/how-ai-is-revolutionising-the-role-of-the-literary-critic.

Mani, Inderjeet and James Pustejovsky (2012). *Interpreting Motion: Grounded Representations for Spatial Language*. New York: Oxford University Press.

Manning, Christopher D., Kevin Clark, John Hewitt, Urvashi Khandelwal, and Omer Levy (2020). "Emergent linguistic structure in artificial neural networks trained by self-supervision". In: *Proceedings of the National Academy of Sciences, vol. 117, no. 48, December 2020*. Pp. 30046–30054. URL: https://www.pnas.org/doi/10.1073/pnas.1907367117.

Martin, Wallace (1986). *Recent Theories of Narrative*. Ithaca: Cornell University Press.

Meister, Clara et al. (2021). "Curveship's automatic narrative variation". In: *Proceedings of the 2021 Conference on Empirical Methods in Natural Language Processing*, pp. 963–980. URL: https://aclanthology.org/2021.emnlp-main.74.pdf.

Mostafazadeh, Nasrin, Nathanael Chambers, et al. (2016). "A corpus and evaluation framework for deeper understanding of commonsense stories". In: *Proceedings of the North American Chapter of the Association for Computational Linguistics: Human Language Technologies (NAACL-HLT'2016)*, pp. 839–849. URL: https://aclanthology.org/N16-1098.pdf.

Piantadosi, S.T. (2014). "Zipf's word frequency law in natural language: A critical review and future directions". In: *Psychonomic Bulletin & Review* 21.5, pp. 1112–1130.

Pinker, Steven (1997). *How the Mind Works*. New York: Norton.

Riedl, Mark (2021). *An Introduction to AI Story Generation*. URL: https://mark-riedl.medium.com/an-introduction-to-ai-story-generation-7f99a450f615.

Rockwell, Geoffrey and Stefan Sinclair (2022). *Hermeneutica: Computer-Assisted Interpretation in the Humanities*. Cambridge, MA: MIT Press.

Ryan, Marie-Laure (2007). "Toward a definition of narrative". In: *The Cambridge Companion to Narrative*. Ed. by David Herman. Cambridge, UK: Cambridge University Press, pp. 22–35.

Sahlgren, M. (2008). "The distributional hypothesis". In: *Ital. J. Linguist* 20.33.

Searle, John R. (1990). "Is the Brain's Mind a Computer Program?" In: *Scientific American*, pp. 26–31.

Shannon, Claude E. and Warren Weaver (1949). *The Mathematical Theory of Communication*. Urbana, Illinois: The University of Illinois Press.

Shipley, K. G. and J. G. McAfee (2024). *Assessment in speech-language pathology: A resource manual*. San Diego: Plural Publishing.

Shklovsky, Viktor (1973). "On the connection between devices of syuzhet construction and general stylistic devices". In: *Russian Formalism*. Ed. by S. Bann and J. E. Bowlt. Edinburgh: Scottish Academic Press. URL: http://dx.doi.org/10.1093/jts/flr173.

Smith, Daniel et al. (2017). "Cooperation and the evolution of hunter-gatherer storytelling". In: *Nat. Commun.* 8, p. 1853.

Taylor, W. L. (1953). "Cloze procedure: a new tool for measuring readability". In: *Journalism Quarterly* 30, pp. 415–433.

Toolan, Michael J. (2009). *Narrative Progression in the Short Story*. Amsterdam: John Benjamins.

Tuckute, Greta, Nancy Kanwisher, and Evelina Fedorenko (2024). "Language in Brains, Minds and Machine". In: *Annual Review of Neuroscience* 479, pp. 277–301. URL: https://doi.org/10.1146/annurev-neuro-120623-101142.

Turing, Alan M. (1950). "Computing Machinery and Intelligence". In: *Mind* 59.236, pp. 433–460.

Venkatraman, Saranya, Nafis Irtiza Tripto, and Dongwon Lee (2025). "CollabStory: Multi-LLM Collaborative Story Generation and Authorship Analysis". In: *Proceedings of NAACL*. URL: https://arxiv.org/abs/2406.12665.

Wang, Yaushian, Hung-Yi Lee, and Yun-Nung Chen (2019). "Tree transformer: Integrating tree structures into self-attention". In: *Proceedings of the Conference on Empirical Methods in Natural Language Processing (EMNLP'2019)*, pp. 1061–1070. URL: https://aclanthology.org/D19-1098/.

References

Wiggins, Geraint A. (2006). "A preliminary framework for description, analysis and comparison of creative systems". In: *Knowledge-Based Systems* 19, pp. 449–458.

Wright, David (2011). "A framework for the ethical impact assessment of information technology". In: *Ethics and information technology* 13, pp. 199–226.

Deep Learning and Transformers   2

> *What we usually consider as impossible are simply engineering problems ... there's no law of physics preventing them.*
>
> —Michio Kaku

2.1 Introduction

This chapter provides a high-level overview of Deep Learning and Transformers. The topic has been discussed in numerous online sources, papers, and textbooks. What is new is my perspective on the narratological and linguistic significance of specific techniques, as well as my own examples and experiments. In making these concepts accessible to students and readers from fields outside computer science, the goal is to keep things as simple and intuitive as possible. The progression here is the standard one found in most textbooks, including (Jurafsky et al. 2024), whose treatment I have adopted.

Accordingly, I start with Simple Linear Perceptrons. I also discuss how to train them. After pointing out their well-known limitations, I then move to Feed Forward Neural Nets, taking the reader on a journey to build a simple net called Tiny Tot. Following a discussion of how to train a net, I explain how the power of the Distributional Semantics introduced in Chap. 1 is brought to bear in numeric representations for word meanings; these are called Word Embeddings. Next, I turn to how rich models of linguistic context are used by Attention mechanisms that allow input words to be influenced by words both nearby and further away. This leads to a discussion of Transformers, the neural net architecture used by state-of-the-art systems like GPT. After explaining how, given a prompt, words are predicted using the Transformer, we describe some practical extensions involving Fine-Tuning and Retrieval-Augmented Generation (RAG), along with some observations on the art of Prompt Engineering.

It is worth noting again that I have made the presentation accessible to students and researchers in the humanities, with starred mathematical sections that they can choose to avoid. This is the only chapter which has quite a few such sections. With this strategy in mind, I hope my account can provide a relatively precise, engaging, and not particularly difficult introduction for students and researchers of all backgrounds to the wizardry or lack thereof in today's Generative AI.

2.2 Simple Linear Perceptrons

2.2.1 SLP Architecture

The Simple Linear Perceptron (SLP for short, and pronounced, say, like 'slurp') was introduced nearly seventy years ago by Rosenblatt (1958). The idea however originates much further back to the work of Gauss in 1795. It is a single neuron model that does some interesting computing. Here is a motivating example.

Consider a simple problem faced by a character in a story. Ali wants to decide whether to propose to Jane, a math major in his college. His ideal mate should be kind and considerate (a character trait called $x1$ for short). If she is kind and considerate, the trait $x1$ will have the value 1, but if she isn't, $x1$ will be zero. She should also (character trait $x2$) love coding as that's what he likes to do. And finally, since working hard also means playing hard, she should hopefully (character trait $x3$) be into thrills like rock-climbing, zip-lining and bungee jumping.

Ali thinks, rather sensibly, that kindness and consideration are crucial. So he gives that character trait a high weight of 90%. We will represent that as a weight $w1$ (corresponding to trait $x1$) of 0.9. As for coding, he gives it a weight $w2 = 0.6$, as he feels it's important, but not a dealbreaker. Finally, we turn to zip-lining and such. Ali gives it a weight ($w3$) of 0.3, because he realizes that thrill-seeking isn't that essential in a companion.

Note that character traits $x1$, $x2$, and $x3$ are have values that are 1 or zero, reflecting their presence or absence, and the weights are real numbers. The inputs to (and outputs from) all neural devices will in fact be numbers, so if we start with a word problem, we have to first convert the problem to numbers. How do we convert text to numbers? The numbers are based on word meanings, from statistics of how the words are used. The details of this conversion from word forms to what are called word embeddings is taken up in Sect. 2.4.

Once Ali gets to know Jane better, he discovers that she is indeed kind and considerate (so for her, the character trait $x1$ has the value 1), but as a star mathematician, she much prefers pen and paper to coding ($x2 = 0$), and for safety reasons, she practices her salsa steps and cheerleading rather than doing extreme sports (trait $x3 = 0$).

Now what should Ali do? He needs to make a go-or-no-go to decision to propose by taking her traits and their weights into account. But Ali also has a gut instinct to propose that can't really be broken down further. On a scale of zero to 1, his gut instinct is 0.7. So he

2.2 Simple Linear Perceptrons

Fig. 2.1 SLP with activation function

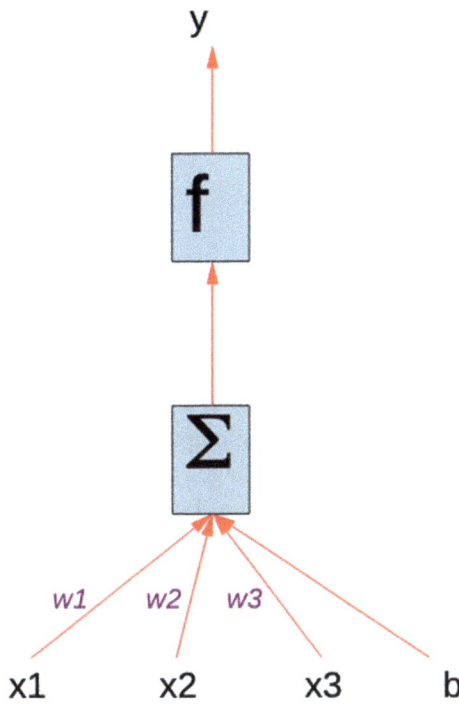

needs to add that in as well. (Technically, this is called a *bias* term b, so we have $b = 0.7$.[1] Ali's overall weighted sum calculation is thus the score: $(1 \times 0.9) + (0 \times 0.6) + (0 \times 0.3) + 0.7 = 1.6$.

But how can he translate the total score to go-no-go? He can use a threshold of zero: if the score is greater than zero, the answer is 'Yes' and he should propose; if not, the answer is 'No'. (The use of a threshold function like that is called a *step function*.) The SLP machine that corresponds to his decision-making is shown in Fig. 2.1. Here, the big $\sum$ sign means a **weighted sum**. The function f, whether in the form of the step function as in the case here, or some other function, is called an **activation function**.

The next section goes into a few mathematical details that can be safely skipped by readers who aren't interested. Literature lovers, however, may note the loose resemblance of Ali's story to that of Graeme Simsion's amusing, and best-selling, novel *The Rosie Project*.

---

[1] The word 'bias' is in fact an odd term to choose for something so important as gut instinct, but it derives from early radio technology where components had to be set, or biased, away from neutral positions to achieve the right operating conditions.

2.2.1.1 Formalizing the SLP*

Let us summarize what we just said in mathematical notation: A basic SLP will compute a weighted sum z (here the n, the length of the input, is 3):

$$z = \sum_{i=1}^{n} w_i x_i + b \qquad (2.1)$$

Let's base the output for the decision based on some function f of the weighted sum z. In the simplest case, the function should return 1 if the decision is 'Go Ahead', and 0 if it's 'No Way'. A trivial function to convert the real number to a boolean is the above-mentioned *step function* $f = step(z)$ whose value is 1 if $z > 0$, and 0 otherwise. The step function makes use of a threshold, in this case 0. So we have $y = f(z) = step(z)$.

In Ali's case, the output $y = 1$ and Ali should seek Jane's hand in marriage.

Suppose Ali changes his mind suddenly and decides to consider Jill instead. We will have different values for $x1$, $x2$, and $x3$, but the weights remain the same. Plugging in the new values, we have a new recommendation for Jill.

Instead of considering $x1$, $x2$, etc. separately, consider the input X as a **vector** of **dimension** d, in this case, $d = 3$. So we have values of X for Jane, Jill, etc. Likewise, the weights W, which are built in, will be gathered in a vector of dimension d. Thus in our little SLP, the dot product $W \cdot X$ is a scalar, as is the weighted sum z, the output y and the bias b. The vector form of Eq. 2.1 is:

$$\begin{aligned} z &= W \cdot X + b \\ y &= \text{step}(z) \end{aligned} \qquad (2.2)$$

In terms of math notation, the dot product (or scalar product) of two vectors $P = \begin{bmatrix} p_1 & \ldots & p_d \end{bmatrix}$ and $Q = \begin{bmatrix} q_1 & \ldots & q_d \end{bmatrix}$ of the same dimension d, written as $P \cdot Q$, is $\sum_{i=1}^{d} p_i \times q_i$. It results in a scalar.

Logical Expressiveness of Eq. 2.2

Note that the SLP in Eq. 2.2, despite just being a single neuron, can handle certain basic logical operations. Let the dimension d of X and W be 2. We can then compute for different values of X (e.g., $x1 = 0$, $x2 = 0$, $x1 = 0$, $x2 = 1$, etc.), functions such as OR (e.g., $W_1 = 1$, $W_2 = 1$, $b = -0.5$) and AND ($W_1 = 1$, $W_2 = 1$, and $b = -1.5$). The reader can try her hand at setting up NOT. These three operations are in fact together rather powerful. They are sufficient for propositional logic, which is the most basic system of logic. Propositional logic (aka the propositional calculus) was developed a little while before the advent of neural nets, in the 3rd Century BCE by the Greek Stoic philosopher Chrysippus of Soli. It also forms the basis, among other applications, for digital circuits, though they aren't implemented by SLPs.

2.2.2 Training a SLP

How do we get the SLP weights? We would like to have the weights be trainable from data rather than be hardwired. We can start with initial weights being random or chosen from some specific distribution, and then adjust those through learning. This is analogous to what Ali himself might go through: as Ali gains in experience, his weights should change accordingly.

Whenever a neural net produces output (which, as we have seen, is in the form of numbers), we have to compare that output against a true answer or ground truth. (In Ali's case, the true answers may be near impossible to find in advance!) The *loss L* is the difference between the network's pre-activation output and the true answer. By pre-activation output, I mean the weighted sum score in Fig. 2.1, rather than the activation step function's go-no-go output.

Let us now turn to the learning method that an SLP can use to train its weights so as to have the least loss.

2.2.2.1 Stochastic Gradient Descent

To train (or teach) a neural network like the SLP in Fig. 2.1 the right weights, it has to be given a training set. This is no more than a set of example inputs along with the true answer for each example. In training, as the network processes each example, it can see how far its output is from the true answer, in other words, it can compute its loss. Let's pretend for the sake of argument that Ali has access to the SLP and he wants the AI to learn to help him with his future decisions, as a sort of matchmaking aid.

Ali starts off with guesswork. He doesn't really know how important each trait should be, so he sets the three weights to 0.5 each, plus a fixed gut instinct bias of 0.7. As training data, he dips into his social network.

He first selects his sister's friend Esmeralda. She's kind and considerate, but not at all into coding, castigating it as a 'nerd' activity. She's also a thrill-seeker and bungee expert, in fact, it was she who first stimulated his interest in extreme sports. Her three character traits are thus $x1 = 1$, $x2 = 0$, and $x3 = 1$. The SLP's weighted sum calculation would be $(1 \times 0.5) + (0 \times 0.5) + (1 \times 0.5) + 0.7 = 1.7$. What is the ground-truth for this example? He knows that her stereotyping of coders as nerds makes her not quite the ideal match, and the reality, or ground truth score, is that she should be rated at 0.8.

The network has overshot and rated Esmeralda too high with an absolute value of the loss (i.e., ignoring any negative sign) being 0.9, so its weights must be lowered. The SLP invokes its Stochastic Gradient Descent (SGD) algorithm to nudge those weights downwards, say, to 0.3 each.

The next training example is Jane. As we know, her character traits are $x1 = 1$, $x2 = 0$, and $x3 = 0$. The weighted sum for Jane would be: $(1 \times 0.3) + (0 \times 0.3) + (0 \times 0.3) + 0.7 = 1.0$. The actual ground truth for Jane is a score of at least 1.8 (to tell the truth, he values her even more than that). So the network has undershot this time, with an absolute value of loss of 0.8. This time, the SGD bumps the weights upwards slightly, say to 0.4.

The reader will get the drift here. Each time a new training example is fed in, the network's weighted sum is compared to the true value, and based on the absolute value of the loss, the weights are nudged lower or higher. Over many training examples, the weights settle to fit the examples. Eventually, the network learns which traits truly matter and by how much, just like Ali refining his knowledge as he makes more friends.

Once the network is fully trained in this way, when a new test example arrives (in the form of a person), her character traits are fed to the trained SLP. The weighted sum is computed and fed to the step function, which makes the go-no-go decision.

To summarize, with each input example, the SGD algorithm computes (in this simplified presentation) the absolute value of loss and updates the weights in a direction that shrinks the loss, aimed at getting a hopefully better result on the next example. The learning rate determines how large an update to take on each example. If it's too large, the weighted sum might overshoot, if it's very small, progress could be at a snail's pace. To speed up learning, examples are also grouped into **mini-batches** (instead of training on just one example at a time).

How long does one keep training the network? One could stop when the loss drops under a particular ceiling value, or after a fixed number of training steps (useful if the loss lingers stubbornly above the ceiling).

> **SGD in a nutshell**
> SGD gently nudges the network's weights so that its output matches the true answers more closely over time. The learning rate determines the size of the update step on each example. The algorithm stops when the loss drops under a particular ceiling, or after a fixed number of training steps.

Because of its efficiency and effectiveness, SGD is widely used in training neural networks.

2.2.2.2 SGD Details*

The Loss L for each input i can be calculated by comparing y_i against the true value T_i. A typical comparison function is Mean-Squared Error (MSE), or $L = (T_i - y_i)^2$.

L will measure how much the NN-generated output differs from the correct output. As it sees each training example X, the neural net will try to adjust its weights to minimize the loss. In other words, we want to minimize, over N training examples:

$$\frac{1}{N} \sum_{i=1}^{N} L(y_i, x_i, w)$$

2.2 Simple Linear Perceptrons

Recall that at each step of the SGD algorithm, we compare the output value to the true value, and then guess how to change the weights so as to move closer to the true label. At that step, the slope of the loss function points in the direction of increasing loss. To minimize the loss function, we move the weights in the *opposite direction of the loss function's slope*. The new value for each weight is towards the opposite direction of the gradient. It is the old value of the weight minus the learning rate η times the gradient of the Loss function L at t. Thus, at step t:

$$w(t+1) \mathrel{-}= \eta \Delta_w L(y_i, x_i, w(t))$$

> **Slope of loss function and learning rate**
> The slope of the loss function is a partial derivative of the loss function with respect to the weights, and it points in the direction of increasing loss. To minimize the loss function, we therefore have to move the weights in the opposite direction. The SGD is efficient in taking only *a random sample x_i* from the training data, rather than computing the loss at step t over all N samples of the input training data. The generic Gradient Descent algorithm does the latter, but is generally less used as it's far less efficient.
>
> The learning rate η is a hyper-parameter that adjusts how fast we move towards the true value. If it's too slow, the training will take too long, and if it's too fast, the algorithm may overshoot the minimum at a given step and then will have to go back. The learning rate is tied to the step of the algorithm, typically starting faster and then slowing down to finely cruise in to a soft-landing close to the true label.

The algorithm terminates when, say, the output is so close to the real value that the loss is acceptable. For that, a tolerable ceiling for the loss is used. When the Loss function is convex, meaning that its graph is cup-shaped—where the line segment between any two points lies on the graph or above it (or nearly so)—and if η decreases at an appropriate rate, the SGD can be shown to converge towards the global minimum. Otherwise, the algorithm can get trapped in a local minimum. As mentioned earlier, in training with large datasets, for safety, in case we never converge below the ceiling, we use as a termination condition a fixed number of steps or batches of input.

2.2.3 SLP Limitations

2.2.3.1 Introduction

An SLP is a simple type of AI model that makes decisions by drawing a line to separate data into two categories. If a data point is on one side of the line, the model gives one answer

'Yes', and 'No' if the point lies on the other side. This dividing line is called the *decision boundary*.

However, not all problems can be solved with just a straight line as a separator. For example, imagine Ali has to choose between marriage or grad school, but not both. That's an example of the XOR (or Exclusive Or) problem, where no single straight line can separate the 'Yes' and 'No' cases. Because of this, the basic SLP fails for XOR problems, and we need more advanced models to handle curves and more complex patterns. To handle complex problems, we need neural networks with multiple layers, which can create more flexible decision boundaries (like curves or other shapes) instead of just straight lines.

2.2.3.2 More on SLP Limitations*

In Fig. 2.1, the SLP activation function is non-linear, as it obviously has a step shape, not a line. Nevertheless, the weighted sum z is a linear combination of its inputs, and this is a straight line. The decision boundary for a classifier like the SLP is a line that divides the data into two classes. Every data point on one side of the line is in one class (output is 1), and every data point on the other side of the line is in another class (output is zero).

An SLP with a non-linear activation function is also known as a logistic regression classifier (LRC). In general, the decision boundary for any classifier is a surface (line, curve, plane, hyperplane, etc.) which divides the data into multiple classes.

Most of the problems that are of interest in this book are not linearly separable, and so an SLP won't work for them. In other words, a line as a decision boundary will not be able to split the data neatly. We therefore need something more than a SLP for even a simple problem like Ali's.

2.3 Multi-layer Neural Nets

We now consider neural networks with multiple layers, or so-called hidden layers. A hidden layer is a layer in between input and output. Having multiple such layers helps solve more complex problems (beginning with XOR) by letting the model bend decision boundaries instead of just drawing straight lines.

Multilayer networks with non-linear activation functions are called **neural nets**. First we consider the most basic kind: a **Feed Forward Neural Net**.

Let us start with an inspiring assistive device called No Negativity. This automaton is a simple function which takes a number as input. If the number is negative, it outputs zero, otherwise it outputs the input number. In other words, it passes along any positive value and blocks negative values. Simple, right?

In reality, No Negativity is called ReLU (for Rectified Linear Unit). Its curve is a bent L shape—flat for negatives and sloping up for positives. If you care to inspect it, it's shown in Fig. 2.2.

$$RelU(r) = max(0, r) \qquad (2.3)$$

2.3 Multi-layer Neural Nets

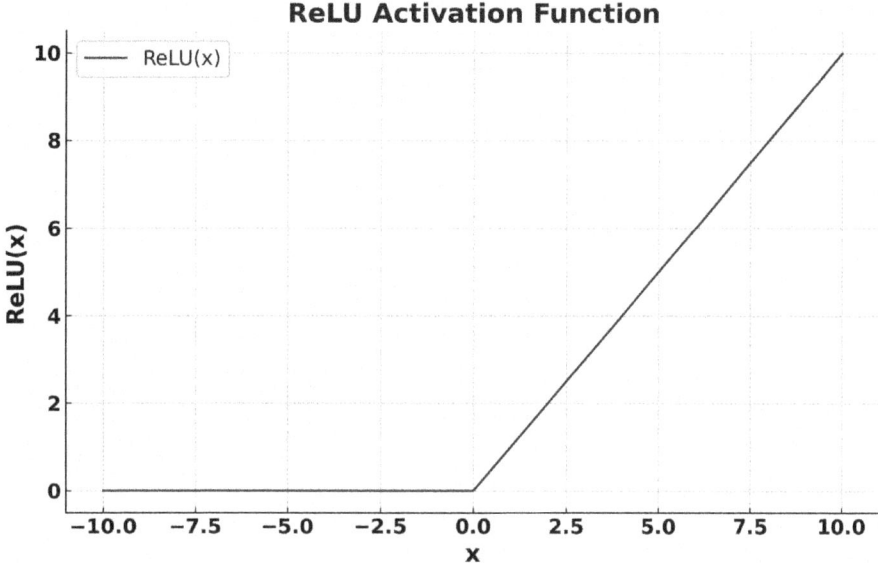

Fig. 2.2 Plot of RelU

> **More on RelU**
> RelU is a seemingly trivial function, as it outputs 0 for negative inputs and returns its input otherwise. It thus has a reversed, slanted L-shaped curve, consisting of two connected lines, a flat horizontal line at zero (for negative inputs), and a diagonal line sloping up from when the inputs turn positive. So though piecewise linear, it's non-linear.

So far, we've seen two rather weirdly-named functions, that nevertheless compute trivial arithmetic. The first was the step function, the second RelU. There is also a third, called Sigmoid. It turns any number into a value between 0 and 1. In fact, small numbers will become close to zero, and large numbers become close to 1. Astonishingly, all the computing devices inside neural nets (the neurons, as it were) are simple arithmetic functions like these! The smarts come in the kinds of connections, the training and inference algorithms, and the vast scale of the net.

2.3.1 FF NN Architecture*

> Adding a single hidden layer with a non-linear activation function addresses the XOR problem, as the reader can verify as an exercise from Fig. 2.3 (adapted from Jurafsky et al. 2024, 3rd. Edn., Fig. 7.6.).

Fig. 2.3 Solving XOR with hidden layer and RelU. Unlabeled links, except for output to y, are all +1.0

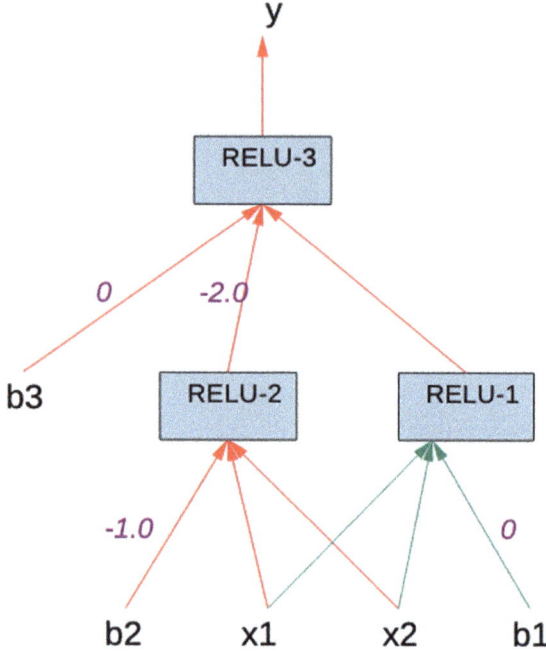

Another non-linear activation function uses the above-mentioned *sigmoid* σ:

$$\sigma(r) = \frac{1}{1 + e^{-r}} \qquad (2.4)$$

Thus, for our SLP output y using sigmoid as the activation function f applied to the weighted sum z, we would have:

$$y = \sigma(W \cdot X + b) = \frac{1}{1 + e^{-(W \cdot X + b)}}$$

More on sigmoid
Sigmoid maps its input into real numbers from 0 to 1, with a sleek and almost S-shaped curve, pushing values closer to the endpoints 0 and 1. Its curve rises slowly from the value zero for large negative inputs through a steep rise at $x = 0$ reaching 1 for large positive inputs (Fig. 2.4).

It's worth noting that hidden layers must have non-linear activations. A multi-layer network with linear activations won't be able to address XOR or other non-linear problems, since a composition of linear functions is just a single linear function, turning such a network back into an SLP.

2.3 Multi-layer Neural Nets

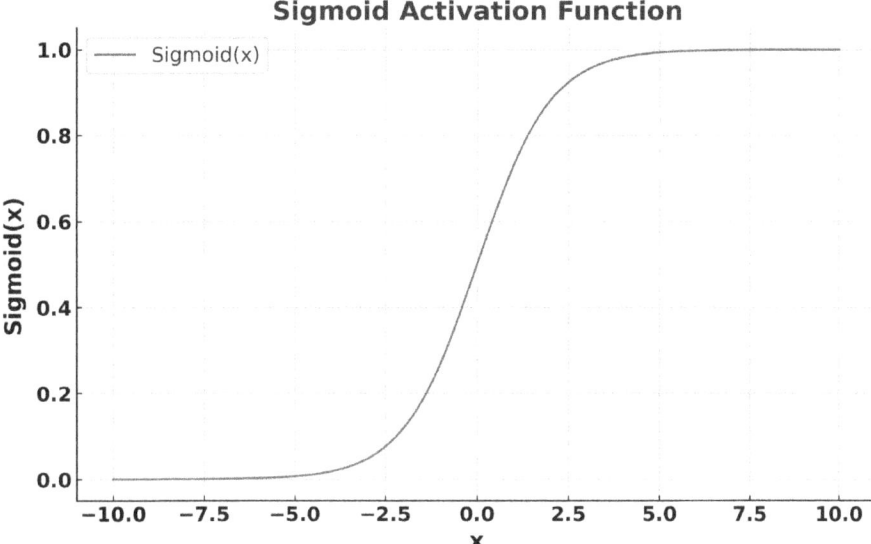

Fig. 2.4 Plot of sigmoid

2.3.2 FF NN Example

2.3.2.1 Introduction

Let's do something really simple. Let's start with a baby FF NN with just two intermediate layers, called Tiny Tot.

Let's call the input of the net X and the output Y. X might be the representation of a quote off the internet, and Y might be whether it's judged acceptable, on a scale of zero to 1. Or X could be the representation of a prompt, and Y could be a score for each possible next word, with that score being a probability between zero and 1. That, rather loosely speaking, is a classic framing for problems that LLM solve. Since neural nets just deal with numbers, X has to be converted to a sequence or array of numbers first.

So, given the prompt "a hilarious", we need to design a FF NN which will give you scores for the next word being 'nightmare', or 'movie', or 'comedy', etc.

2.3.2.2 Tiny Tot Setup

Each layer will have weights and a bias. Let W1 be the weight of layer 1, and it will have a bias B1. Layer 1 will, like any decent layer, have an output, and let's call it Q1. So layer 1 maps between input X and output Q1.

Now we come to layer 2. We have to pass the output Q2 as input to layer 2. Layer 2 will itself have a weight W2 and bias B2, and will have an output Q2.

Finally, we will let Y, the final output, be Q2.

That's it, except that we haven't discussed how Q1 is produced given X, W1, and B1. This means we have to use some sort of function to map from X, W1, and B1 to Q1. This is not that scary, so let's stay positive. In fact, let's use our old friend No Negativity! Its technical name is of course RelU, the famous blocker of negative values. So RelU will take X, W1, and B1 to produce Q1.

We have now together built the first layer of our baby Tiny Tot. Now we have to take Q1, W2, and B2 to produce Q2, i.e., Y. Remember that Y will have to represent scores for different possible words that follow. The way to do this is for Y to be a probability between zero and 1, so that probabilities of all the next word alternatives sum to 1.

Fortunately, there is a handy device at hand to do just that, called *softmax*. It takes a set of input weights and converts them into probabilities, ensuring that all probabilities sum to 1. The second layer is also done, thank goodness!

Tiny Tot is shown graphically in Fig. 2.5. Here we use square brackets Q[1] for Q1, etc., but don't worry about that. Note that since Y is the same as Q2, we don't want X to feel bad, so for symmetry we make X the same as Q0.

2.3.2.3 Next Word Prediction with Tiny Tot

Now, how do we use this baby net to guess the next word in a sentence? Consider the prompt "A hilarious". We need a numeric input representation X for it, say the array of two numbers 0.2 and 0.4. Now let us see how that input X passes through the two layers. Each layer multiplies the input by weights and adds in the bias to produce the weighted sum, which

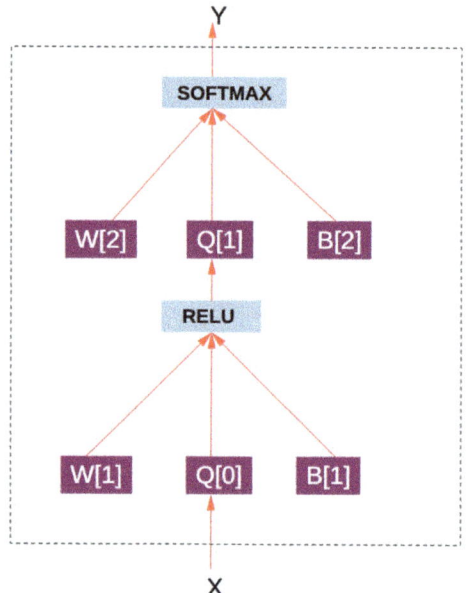

Fig. 2.5 The 2-layer FF NN Tiny Tot

is fed to an activation function. Tiny Tot's output Y will be the two probabilities indicating how likely each next word ('moment' or 'movie') is.

In the first layer, Tiny Tot multiplies the input X, which is 0.2 and 0.4, by this layer's weights and biases to yield the weighted sum. That will be fed through the RelU activation function, which as the reader may recall is No Negativity, leaving positive numbers untouched and zapping negative numbers to zero. The output of the layer will also be two numbers, corresponding to Q[1] in Fig. 2.5.

Let's zoom in a bit. The first layer has four weights: 0.1, 0.2, 0.3, and 0.4, plus two small biases of 0.1 each. So Tiny Tot takes the first number in X, which is 0.2, multiplies it by weight 0.1, and then the second number in X, which is 0.4, and multiplies it by weight 0.2, adds those results together, then adds the bias 0.1. That comes out to 0.18. For the second output, it does the same thing but uses 0.3 and 0.4 as the weight multipliers, ending up with 0.34. These results are both positive, so RelU leaves 0.18 and 0.34 unchanged as the output Q[1] of the layer.

In the second layer, it does the same kind of arithmetic again on Q1, which is the array of two numbers 0.18 and 0.34, using the same four weights and offsets, resulting in the array the numbers 0.186 and 0.29.

Finally, Tiny Tot applies softmax, which automagically converts these two numbers into two probabilities that sum to 1: 0.474 and 0.526, which is the final output Y of the network. Because 0.526 is bigger than 0.474, Tiny Tot predicts 'movie' is a somewhat more likely to follow "A hilarious" than 'moment'. In other words, the net concludes that "A hilarious movie" is more probable than "A hilarious moment".

This example also illustrates that the differences in probability don't have to be that large to select the most likely next word. Finally, it also makes clear how simple the underlying arithmetic calculations really are. As we shall see, it is amazing that neural nets built on similar simple mechanisms can do some of the astonishing things that they have come to be known for.

The next section details how the weights are computed, for those who want to embrace the math. However, it can be safely skipped by those who don't.

2.3.3 FF NN Example: Computational Details*

To refer to a particular layer, we will avail of square brackets, so instead of W1 and W2, we will use $W[i]$ to refer to the weight matrix at layer i, where $i = 1$ is the first hidden layer. Thus, a 2-layer FF NN will have one hidden layer. $W[i]$ (where $i >= 1$) will have dimension $d[i-1] \times d[i]$. $Z[i]$ will be the weighted sum vector at layer i. We use Q to refer to the output of each layer, once the activation function $f[i]$ has been applied to the weighted sum $Z[i]$. (In terms of notation, we overload the function $f[i]$, to map the activation function $f[i]$ across each element of its input. Thus while the original function applied to each element returns a scalar, the overloaded function returns a vector.) The input vector X of dimension $d[0]$ will be referred to as $Q[0]$. $Q[1]$ is of dimension $d[1]$, which may not be the same as $d[0]$. Y is the final output vector, of dimension $d[n]$.

The value for each element of the vector $Z[2]$ is a scalar number. To convert $Z[2]$ to a probability distribution, we need to make all output values in the network output range from zero to 1 and also sum to 1. For this we use our simple but crucial normalization function *softmax*. Recall that it takes a set of input weights and converts them into probabilities, ensuring that all probabilities sum to 1. It does that by taking each output value, exponentiating it and then normalizing by the sum of all the exponentiated values. It returns a vector of the same dimension as the input $Z[2]$. In general, then, given that Z is of dimension d:

$$\text{softmax}(Z_i, Z) = \frac{e^{Z_i}}{\sum_{j=1}^{d} e^{Z_j}} \tag{2.5}$$

Thus if the final output vector Y was of dim 2, we could have $Y = \begin{bmatrix} 0.47 & 0.53 \end{bmatrix}$, which would represent a probability distribution of two classes, with the second class being more likely.

Let's define the functions inside the 2-layer FF NN Tiny Tot. Before diving in, we need to learn how to multiply matrices. Let A be a $m \times n$ matrix and B be a nxp matrix. Then $A \times B$ is the matrix C whose entry c_{ij} in the ith row and jth column is the ith row of A (call it A_i) times the jth column of B (call it B_j). So, $c_{ij} = a_{i1}b_{1j} + a_{i2}b_{2j} + \ldots + a_{in}b_{nj}$.

Example 2.1

$B[1] = B[2] = \begin{bmatrix} 0.1 & 0.1 \end{bmatrix}$
$f[1] = \text{RelU}$
$f[2] = \text{softmax}$
$Q[0] = X$
$Z[1] = W[1] \times Q[0] + B[1]$
$Q[1] = f[1](Z[1])$
$Z[2] = W[2] \times Q[1] + B[2]$
$Q[2] = f[2](Z[2])$
$Y = Q[2]$

Universal Approximation Theorem

The *universal approximation theorem* states that a neural network with just one hidden layer can approximate continuous functions with any desired precision. (These are functions over compact sets, where a set is compact if a finite number of open covers can cover the entire set.) This theorem implies that most NLP tasks can be approximated using neural nets, in principle. It's ultimately a matter of designing the right net.

2.3 Multi-layer Neural Nets

Say the input X is a weight vector for the words "A hilarious", e.g., $X = \begin{bmatrix} 0.2 & 0.4 \end{bmatrix}$. So X is of dimension 2. The output Y will a probability distribution over all possible words that could follow. We will just restrict it to a distribution of two words, 'moment', and 'movie'. Y will also be of dimension 2. For each layer i, $Q[i]$ will be of dimension 2, and $W[i]$ will be of dimension 2×2.

Assume that Tiny Tot from Example 2.1 is trained with the following weights:

$$W[1] = W[2] = \begin{bmatrix} 0.1 & 0.2 \\ 0.3 & 0.4 \end{bmatrix}$$

$$B[1] = B[2] = \begin{bmatrix} 0.1 & 0.1 \end{bmatrix}$$

To run a trained n-layer net where the weight matrices W have all been computed, we use the **Forward Pass** method in Algorithm 1.

Data: input vector X, max layers n
Result: output vector Y
$Q[0] = X$;
for i in $1 \ldots n$ **do**
$\quad Z[i] = W[i] \times Q[i-1] + B[i]$;
$\quad Q[i] = f[i](Z[i])$;
end
$Y = Q[n]$;

Algorithm 1: Forward Pass

Running *Forward Pass* on the trained Tiny Tot net gives us the output in Example 2.2.

Example 2.2

$Q[0] = \begin{bmatrix} 0.2 & 0.4 \end{bmatrix}$
$Z[1] = W[1] \times Q[0] + B[1]$
$= \begin{bmatrix} 0.1 & 0.2 \\ 0.3 & 0.4 \end{bmatrix} \times \begin{bmatrix} 0.2 & 0.4 \end{bmatrix} + \begin{bmatrix} 0.1 & 0.1 \end{bmatrix} = \begin{bmatrix} 0.18 & 0.34 \end{bmatrix}$
$Q[1] = f[1](Z[1]) = \begin{bmatrix} max(0, 0.18) & max(0, 0.34) \end{bmatrix} = \begin{bmatrix} 0.18 & 0.34 \end{bmatrix}$
$Z[2] = W[2] \times Q[1] + B[2]$
$= \begin{bmatrix} 0.1 & 0.2 \\ 0.3 & 0.4 \end{bmatrix} \times \begin{bmatrix} 0.18 & 0.34 \end{bmatrix} + \begin{bmatrix} 0.1 & 0.1 \end{bmatrix} = \begin{bmatrix} 0.186 & 0.29 \end{bmatrix}$
$Q[2] = f[2](Z[2]) = softmax(0.186, 0.29) = \begin{bmatrix} 0.474 & 0.526 \end{bmatrix}$
$Y = Q[2] = \begin{bmatrix} 0.474 & 0.526 \end{bmatrix}$

Thus, the inference-time *Forward Pass* calculation predicts that after seeing "A hilarious", the next word should be 'moment' with probability 0.47 and 'movie' with probability 0.53. So according to Tiny Tot, 'movie' is somewhat more likely.

2.3.4 What's Wrong with Tiny Tot?

Alas, our little Tiny Tot is not really well-equipped as an LLM, let alone to handle NLP. Here's why:

1. It takes only two words as input X. This of course can be extended.
2. The input weights will have to be acquired somehow, to accurately represent the meanings and distributional properties of these words. We will discuss how to do that shortly, in Sect. 2.4.
3. The weights W and B are hardwired. These can be acquired from training, as we will see next, in Sect. 2.3.5.
4. The more serious problem, however, is that the Tiny Tot architecture is too tiny and simplistic to compute, even with training, appropriate probabilities for the next word. It doesn't even have a Language Model for predicting the next word. And if it were to have one, how rich a model of context should we use, since bigrams aren't, as we know, sufficient? The answer lies in the Self-attention mechanisms explained in Sect. 2.5.

2.3.5 Training a FF NN

2.3.5.1 Stochastic Gradient Descent for FF NN

We have already seen, in Sect. 2.2.2.1, how SGD works to train Simple Linear Perceptrons (SLPs). Remember: this is done by gently nudging the network's weights, as in our friend Ali's case, so that its output matches the true answers more closely over time. The reader will therefore have already suspected that something similar could be used to train FF NNs. And the reader will have guessed right.

However, given that the FF NN has multiple layers, the error-nudging SGD algorithm has to be adapted accordingly. That's because the update in the SLP's SGD algorithm is only for the weights for the final layer, where we can actually compare the current output value with the true value. What about the updates of the weights of intervening (aka hidden) layers before the final net output has been reached?

Backpropagation (or Backprop for short) is the answer. After the network produces an output, we compare it to the correct answer and measure the Loss. This difference gets translated into an error signal which is then passed backward through the network, from the final layer, through each hidden layer, and back to the start.

To understand this more easily, let's turn to our favorite food—chocolate. Imagine an assembly line in a chocolate factory. Assume the quality control inspector at the end tastes the chocolate and finds it to be too bitter (analogous to a high Loss in comparing the final output and the correct answer). The inspector (who has the ideal day job for some readers, including myself) travels back through each station (i.e., each earlier layer in the network), and finds whether and how that station added to the bitterness (too much dark chocolate,

2.3 Multi-layer Neural Nets

for example, or too little sugar), and then instructs the station how to make small fixes. By repeating this process many times, the network (like the factory) fine-tunes its steps so the end result is closer to the true answer.

When networks have many layers, as modern nets do, the error signals can sometimes become very weak by the time they reach the earliest layers. This is called, for reasons that are clarified in the next section, the *vanishing gradient* problem. Staying positive, by using ReLU activation functions, help keep these signals stronger. Another challenge is making the network generalize better to new test examples that are different from the training examples, rather than *overfitting* to the training data (and performing badly on such new test examples). One solution here is to turn off some connections temporarily so the network can generalize better. The latter method is referred to rather pejoratively as *dropout*.

If you understood the intuition behind Backprop, and want more mathematical details of how SGD uses it, feel free to read the next two sections.

2.3.5.2 The Loss Function: Details*

To train, we need, as with SLPs, to be given the network structure, namely, the number of layers n, the dimensions of X, and the dimensions of each $W[i]$ for each layer i in $1\ldots n$. And as with SLPs, we also need a **Loss function** L which the training algorithm will minimize.

Recall that our trained NN in Example 2.2 output a final vector Y, which represented the probability distribution predicting 'moment' with probability 0.47 and 'movie' with probability 0.53. Using the standard machine learning notation $\hat{y}$ to represent the output of a model, $\hat{y}$ is what we have been calling Y in Fig. 2.5. So we have $\hat{y}_1 = 0.47$, and $\hat{y}_2 = 0.53$. The number of classes $c = 2$. The NN's predicted distribution has to be compared against the true labels y, namely a one-hot vector with the value 1 in the position for the correct label and the value zero elsewhere. Assuming 'movie' is the correct label, this could be:

$$y = \begin{bmatrix} 0 & 1 \end{bmatrix}$$

Thus we have $y_1 = 0$, and $y_2 = 1$.

We can now define, a fundamental metric of LLMs, the **Cross-Entropy Loss**, shown in Eq. 2.6:

$$L_{CE}(y, \hat{y}) = -\sum_{i=1}^{c} y_i \log \hat{y}_i \tag{2.6}$$

Since y is one-hot, it's only a single y_i in position k (here $k = 2$) where y_i is 1 and the rest of the y_is are 0. So, Eq. 2.6 simplifies to Eq. 2.7:

$$L_{CE}(y, \hat{y}) = -\ln \hat{y}_k \tag{2.7}$$

For efficient computation, here we use, as in most machine learning applications, logs to the base e, rather than to the base 2.

When the model assigns probability of 1 to the correct class and zero to the rest, $L_{CE} = -ln(1) = 0$, so perfect prediction means a cross-entropy loss of zero. The closer the loss is to zero, the better. In our example:

$$L_{CE}(y, \hat{y}) = -ln\ \hat{y}_2$$
$$= -ln\ 0.53 = 0.63$$

Given that the perplexity with natural logs is $e^{L_{CE}}$, the perplexity of this model is 1.88.

2.3.5.3 SGD for FF NN: Details*

How do we compute $L_{CE}(y, \hat{y})$ for our FF NN in Example 2.1?

Assume that the true next word is 'movie'; The initial loss given the initial weights is $-[1\ ln(0.46) + 0\ ln(0.54)] = 0.776$. So, we must now update the weights. While the above formulation of SGD is sufficient for training a single-layer NN with non-linear activation function, it is not sufficient for a FF NN which has hidden layers. as we have seen.

What we need is to extend the SGD algorithm with a Backward Pass. Once we compute, at a given iteration, the loss function of the output from the forward pass, we then calculate the gradient of the loss function for that output with respect to the original input. If there is no layer in between, that's as in the SGD for the SLP. But if there is a layer in between, we use the chain rule for differentiation. This, as we mentioned earlier, is called Backprop.

Instead of using the Loss function in Backprop, we can simplify matters by computing an *Error* at each layer, which is simply the difference between the current output and the true labels vector (let us call it T):

$$Error[i] = Q[i] - T \qquad (2.8)$$

However, we may still want to preserve the Loss function for reporting and diagnostics.

We are now ready to present the SGD algorithm for FF NN. Here θ represents the parameters of the network. L is the Loss function. X is all the training data, X_i being a current input sample. T is the training vector of true labels. ϵ is the error ceiling, η is the learning rate. The number of layers is given by n, and MaxEpochs controls termination. The details are shown in the last detailed algorithm in this book, Algorithm 2:

2.3 Multi-layer Neural Nets

Data: θ, loss L, input X, true labels T, max layers n
Result: θ
1. Initialize weights and biases in θ to small random values ;
2. **repeat**
    2.1 Randomly shuffle dataset X, T ;
    2.2 Randomly pick a sample X_i ;
    2.3 $Q[n] = ForwardPass(X)$;
    2.4 *Compute Loss*: $L = L_{CE}(Q[n], T_i)$;
    2.5 *Backprop*: ;
    $Error[n] = Q[n] - T_i$;
    for *layer r in $n \ldots 1$* **do**
        $\frac{\partial L}{\partial W[r]} = Error[r] \times Q[r-1]^T$;
        $\frac{\partial L}{\partial B[r]} = Error[r]$;
        if $r > 1$ **then**
            $Error[r-1] = (W[r]^T \times Error[r]) \odot f[r]'(Z[r-1])$
        end
    end
    2.6 *Update Parameters*: ;
    for *layer r in $1 \ldots n$* **do**
        $W[r] \mathrel{-}= \eta \frac{\partial L}{\partial W[r]}$;
        $B[r] \mathrel{-}= \eta \frac{\partial L}{\partial B[r]}$
    end
until *Terminate?(MaxEpochs, ϵ)*;

Algorithm 2: Stochastic Gradient Descent for FF NN

Some more details of the mathematical notation: the Hadamard product $\odot$ of two matrices A and B of the same dimensions is the matrix C, each of whose elements is the product of the corresponding elements in A and B. The transpose of a matrix M, written as M^T, is simply M with its rows and columns interchanged. We use the transpose to prep the matrix multiplications, which require that the first matrix have the same number of columns as the second matrix.

Note that in Algorithm 2, $f[r]'$ is the partial derivative of the activation function at layer r.[2]

---

[2] Note also that for a mini-batch, the bias gradient will instead be the sum or mean of Error[r] over the batch dimension, depending on, respectively, whether gradients are to scale with batch size or remain normalized.

> **SGD Efficiency**
> The gradient computation in back propagation requires many multiplications, as the chains can drag out to very long in a large network. These multiplications can shrink the gradients close to zero, called, as we have seen, the vanishing gradient problem. The use of sigmoid can exacerbate the problem as its derivative is a bell-shaped curve with the values squished closer to zero. Vanishing gradients can be partially relieved by using RelU instead, whose derivative is more like a somewhat flattened S-shape.
> To make training even more efficient, instead of each training step taking in a single random vector from a set of all input vectors, we train on mini-batches of a given size. In addition to the usual hyper-parameters like batch size and learning rate, and initialization values for weights, there are also other parameters related to regularization methods that are used to avoid overfitting, like dropout (i.e., dropping certain units or connections), or adding a penalty for large weights.

2.4 Word Embeddings

Let us now put the Distributional Semantics discussed in Chap. 1 to work. Word embeddings are representations of words as arrays of numbers. This allows words to be compared for similarity in that space, where words like 'aim' and 'fire', which occur in similar contexts, would be closer to each other. I emphasize similar contexts, rather than co-occurrence: word embeddings are very powerful because, being an instantiation of Distributional Semantics, they abstract over different contexts of use.

2.4.1 Word2vec

2.4.1.1 Introduction

The starting point for building word embeddings is a relatively simple algorithm called *Word2vec*, developed by Mikolov et al. (2013).

Word2vec is a clever way to teach nets the meaning of words without human supervision. It works by looking at how words appear together in sentences. For example, consider the sentence in Example 1.3, repeated here:

Example 2.3
A hunter to the core, he crouched low, aimed and ...

2.4 Word Embeddings

Let's call this the Hunter sentence. What does Word2vec do with it? The program treats each word as a *target* and looks at nearby words as *context*. It slides a window over the text and, for each target word, tries to predict the surrounding context words. It thus could learn that words like 'hunter' and 'aimed' tend to appear near each other, suggesting they're related in meaning. It creates a mathematical representation (an *embedding*) for each word that captures these relationships.

Embeddings are not just flat arrays of numbers like we were considering for inputs (like X) to our young friend Tiny Tot. Embeddings have a large number of dimensions; each dimension of a word's embedding, i.e., its meaning representation, is an array of numbers, and there may be thousands of dimensions!

One neat feature of Word2vec is that it generates a map where similar words cluster together. Words that appear in similar contexts end up close to each other in this mind-bogglingly multi-dimensional mathematical space. For example, you can perform calculations like this:

$$King - Man + Woman = Queen \tag{2.9}$$

Equation 2.9 shows that the system has learned meaningful semantic relationships between concepts.

These word embeddings have become fundamental building blocks for many language understanding tasks, allowing machines to grasp key distributional facts about word meanings without human supervision.

2.4.1.2 Word2vec Details*

As implied earlier, Word2vec uses a classifier that predicts, given a target word and a context word, the probability that the context word is strongly associated with the target word. The learned probabilities are the values for the embedding. The approach smartly dispenses with any need for supervision during training. It treats context word occurrences near the target word as *positive examples* for training, randomly mixing in negative examples.

Here is an overview of the skip-gram version of Word2vec at work. Consider an input sentence, successively considering one word being the target and the other L neighboring words being the context words.

Let us continue with the Hunter sentence. Let each word x have an embedding vector **x** of fixed length d_m.

The probability of a context word c being positive given that word and a target word w is (with σ being the sigmoid function):

$$P(c = pos|w, c) = \sigma(\mathbf{c} \cdot \mathbf{w})$$

The probability of c being negative is $1 - P(c = pos|w, c)$, which reduces to $\sigma(-\mathbf{c} \cdot \mathbf{w})$.

The cross-entropy loss function L_{CE} will maximize the probability of the target word given true contextual words. Likewise, it will minimize the probability of the target word given the k negative non-neighbor words. It is shown in Eq. 2.10.

$$\begin{aligned} L_{CE}(w,c) \\ &= -\log[P(c=pos|w,c) \prod_{i=1}^{k} P(c_i = neg|w,c_i)] \\ &= -[\log \sigma(\mathbf{w} \cdot \mathbf{c}) + \sum_{i=1}^{k} \log \sigma(-\mathbf{w} \cdot \mathbf{c_i})] \end{aligned} \quad (2.10)$$

Starting with random initial values for word vectors, we then use the (by now intimately familiar!) SGD algorithm to minimize the Loss. The result of running this algorithm is that words that occur in similar contexts move closer in vector space.

> **Semantic properties of** *word2vec*
> Since Word2Vec is not a deep neural net, its vectors are less opaque and exhibit various semantic properties, including reasoning with analogy in the vector space of word embeddings. Thus, additions and subtractions of word embedding vectors give us intuitive results, as in Eq. 2.9 for the semantic relationships of Queen and King.

2.4.2 GloVe

GloVe (Global Vectors for Word Representation) is a word embedding technique developed by Pennington et al. (2017). Unlike Word2vec, it looks at global data reflecting how words appear together across an entire training dataset. For example, consider the words 'ice' and 'steam'. The approach examines those words' co-occurrence probabilities with various probe words, like the word 'solid'. Consider the probability of 'ice' and the probe word 'solid' occurring together in some context, and likewise the rather low probability of 'steam' and the probe 'solid' occurring together. We would expect the ratio of the former to the latter to be large. Similarly, consider the probe word 'gas'. Here we would expect the ratio to be small, since 'ice' would have a low probability of co-occurring with 'gas' whereas 'steam' would have a higher probability. For the probe 'fashion', which is likely equally unrelated to 'ice' and 'steam', the ratio should be close to one. Using such ratios of probabilities, rather than probabilities alone, the approach is better able to distinguish relevant words from irrelevant ones, and is also better at discriminating between relevant words.

The technical details of GloVe's objective function are somewhat involved, and I will skip it here; the interested reader can refer to Pennington et al. (2017). Due to its combining

2.4 Word Embeddings

local context with global co-occurrence information, GloVe works better than Word2vec on many benchmarks. Although Word2vec may indirectly discover across all its training examples that 'steam' co-occurs with 'gas' more than 'solid', it does not directly make use of ratios of co-occurrence probabilities with other words, while GloVe does. And though it is somewhat slower to compute than Word2vec, GloVe exhibits richer semantic similarity, in turn reflected in better word similarity judgments.

2.4.3 GPT Embeddings

While Word2vec and GloVe are basic methods for generating embeddings, contemporary distributional models create multiple vectors for each word reflecting the word's different usage contexts, thereby providing a better handling on polysemy. They are also trained using Transformers. They are thus neural in nature and somewhat more opaque.

Figure 2.6 shows embeddings for some of the words in the Hunter sentence, together with 'man' and several random words. The embeddings were originally in 1536 mind-boggling dimensions, which I thankfully reduced to two. (I generated these embeddings using the GPT model 'text-embedding-3-small'. I then ran Principal Components Analysis (PCA) to

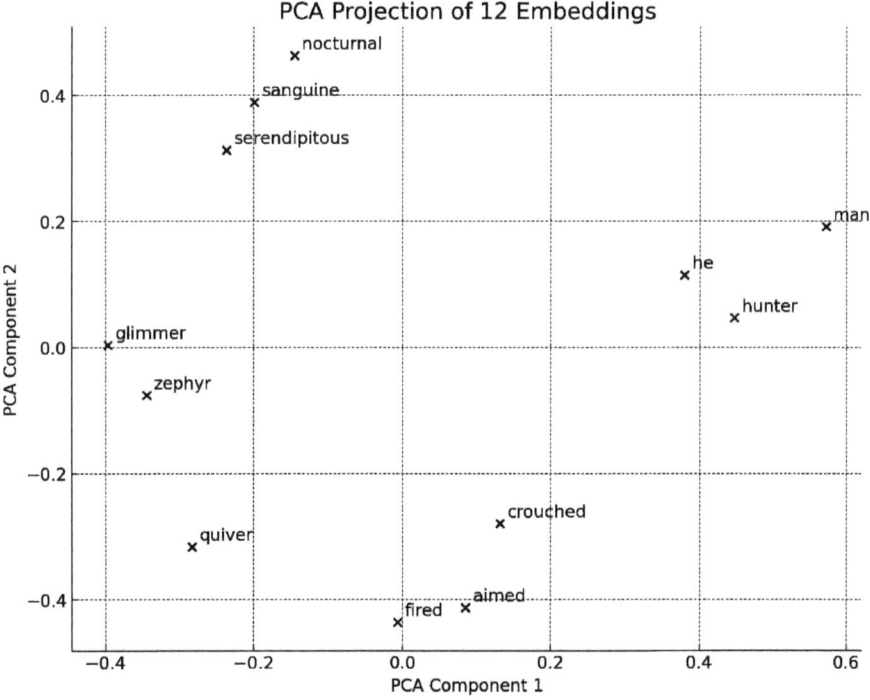

Fig. 2.6 Word embeddings in 2D

automagically reduce to two dimensions.) It is clear that the embeddings for 'aimed' and 'fired' are very close together, with 'he', 'hunter', and 'man' also being somewhat close.

2.5 Self-attention

2.5.1 Background

Consider again the Hunter sentence. Given the prompt that is the token sequence of that sentence, GPT-4o's answers to the question about the hunter are in Example 2.4.

Example 2.4

> **User Prompt**: *A hunter to the core, he crouched low, aimed and . . . .*
> What can you tell me about the hunter? Limit your response to 40 words.
> **Machine**: The hunter is focused, calculating, and precise. His crouching signals readiness, honed instincts, and mastery of his craft. Patient yet poised for action, he embodies a deep connection to the pursuit, waiting for the perfect moment to strike.

However, every sentence has its own context. Consider a case where the Hunter sentence is preceded by the following sentence, in Example 2.5:

Example 2.5
The bullet, he knew, would head straight to the candidate's pathetic little head.

Let's call this the Candidate sentence. Clearly, the word embeddings from language at large may not be appropriate or sufficient for this particular use of 'hunter'. We would like a system that understands narrative to grasp that the hunter is an assassin. That is in fact what GPT does when given the Candidate sentence as additional context. Its verbose answer is as follows:

Example 2.6

> The hunter is deeply contemptuous of his target, viewing the candidate as weak and insignificant. His cold, methodical demeanor suggests a professional assassin or a deeply motivated individual with a personal vendetta. The phrase "pathetic little head" reveals disdain, indicating that the act is not only about precision but also fueled by scorn. His confidence in the bullet's trajectory underscores a ruthless, calculating nature, willing to kill without hesitation or moral conflict.

2.5 Self-attention

The idea of self-attention (Vaswani 2017), in the paper rather cleverly titled "Attention is all you need" (a reference to an old Beatles song), is to have the net pay attention to each given input word in the context of its use in the current sentence or discourse. This extends to long-distance dependencies within and across sentences. Within a self-attention layer, the current word in focus, called the *query*, is compared against the other words, each of which is the *key*. The result of the comparison is the attention *value* for the focus word, namely the attention given to the focus word in context of the keys. Thus, in the Candidate sentence and the Hunter sentence, 'hunter' as a query will be compared against all the words in the sentence (including itself) and in any other prior context as keys.

The relatively simple mechanism of self-attention allows all the query-key computations to be carried out in parallel, since they don't depend on each other. (A hallmark of many important innovations is their essential simplicity!) This fine-grained parallelism makes for much faster and far more efficient processing than NN architectures that are recurrent, where the computation at each time step is dependent on remembering the computation at previous time steps. At the scale at which these models operate, such efficiency gains are invaluable.

The computational details are discussed next, and are optional.

2.5.2 Computing Self-attention*

Let's examine some of the details of the (relatively simple) matrix multiplications (in fact, multiplication is all we need), paying attention to the matrix dimensions to make sure that everything is lined up properly. Let us assume the vocabulary vector Voc of all the words in the data is a one-hot encoding vector of dimension d_{Voc}. So in the Hunter sentence, 'he' will be in some position p in Voc, and $\text{Voc}_p = 1$, and likewise 'hunter' in position q will have $\text{Voc}_q = 1$. The tokenizer yields token ids, which are indices into Voc.

The input representation for each word i in the Hunter sentence, based on the embeddings from language at large, is a vector of dimension d_m, where $d_m \ll d_{\text{Voc}}$.

Let d_m be the dimension of the word embedding. For input position i, the query vector q_i (of dimension d_q) is computed by multiplying a query weight matrix W^Q (of dimension $d_q \times d_m$) with the vector for x_i of dimension d_m. Likewise, the key vector k_i (of dimension d_q) is computed by multiplying a key weight matrix W^K (of dimension $d_q \times d_m$) with the vector for x_i. The value vector v_i (of dimension d_v) is computed by multiplying a value weight matrix W^V (of dimension $d_v \times d_m$) with x_i.

$$
\begin{aligned}
q_i &= x_i \times W^Q \\
k_i &= x_i \times W^K \\
v_i &= x_i \times W^V
\end{aligned}
\quad (2.11)
$$

The query vector q_i and key vector k_j, which are of identical dimension d_q, are compared for similarity, which is computed by a dot product of $q_i^T \cdot k_j$, returning a scalar. (Recall that transposing a matrix is done here to make multiplications compatible.) The attention weight between query i and key j is then a *softmax* normalization of the similarity score. In order to scale down otherwise over-large numerical values and to avoid vanishing gradients, we normalize this scalar value by the square root of the embedding dimension length.

$$\alpha_{ij} = \text{softmax}(\frac{q_i^T \cdot k_j}{\sqrt{d_q}}) \tag{2.12}$$

So far, we have only considered query q_i and the matched key k_j for the attention weight α_{ij}. What happens to the value vector v_j? The self-attention output a_i for position i computes a weighted sum, over all word positions j, of the attention weight α_{ij} multiplied with the value vector v_j:

$$a_i = \sum_{j=1}^{n} \alpha_{ij} \times v_j \tag{2.13}$$

In applications where only the left input up to i must be considered, we use $m \le i$ in the sum instead of n.

2.5.3 A Detailed Example

As the writer Henry Miller once observed, "The moment one gives close attention to any thing, even a blade of grass, it becomes a mysterious, awesome, indescribably magnificent world in itself". With that in mind, let's look at the result of self-attention applied to the Hunter sentence. I tried out the LLM model 'GPT-Neo 1.3B', averaging its attention across heads (see the discussion of multi-head attention below) and then plugged the data into a visualization using heat maps. The result is shown in Fig. 2.7.

The figure shows three matrices displaying snapshots of three network layers (the initial layer, a middle layer, and the final layer, respectively). Each matrix entry reveals how much attention an attending word or *query* (shown on the Y-axis) is paying to the *key* word attended to (on the X-axis). Here the entry values are probabilities expressed as percentages.)

For each attending query word, let us examine the strongest attention score to another attended-to word (x-axis), other than the first word in the sentence. If that connection is much stronger than the attention to all other words (other than itself), that indicates that the query word is pulling relevant contextual information from the key word.

In the initial layer, the attending query 'hunter' matches with strength 11.96 against the attended-to key 'A' to perhaps prep a determiner-noun relationship, which could cue a noun phrase.

In the middle layer (12), the attending queries 'core' and 'he' match to the attended-to key 'hunter' (7.10 and 7.08, respectively) perhaps prepping an anaphoric relation. Likewise, 'aimed' and 'crouched' as queries match to the key 'hunter' (7.15 for 'aimed' and 6.99 and 7.13 for 'crouched'), strongly suggesting subject-verb relations. The query 'core' matches to the key 'the' (7.19), prepping another determiner-noun relation.

2.5 Self-attention

Fig. 2.7 Self-attention for Hunter sentence

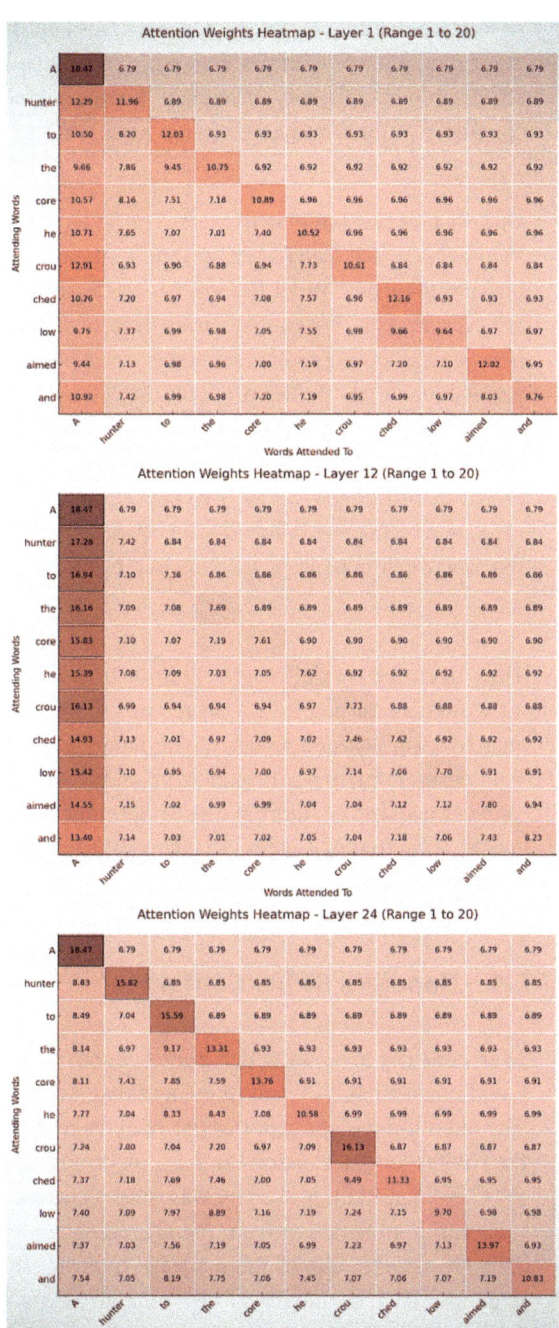

In the final layer (24), query 'the' matches key 'to' (9.17) perhaps to perhaps prepping a prepositional phrase, which is also perhaps prepped by query 'core' matching the key 'to' (7.85). Both parts of query 'crouched' keep a subject-verb relation with key 'he' (7.09 and 7.05). Query 'aimed' has a relation of strength 7.23 to key 'crouch'.

> **Opacity and Self-attention**
> While this sort of data inspection and analysis is interesting, there is a degree of cherry-picking involved. What we need is some sort of explanatory layer or translation of what's going on. We can in certain cases measure objectively, as we noted in Chap. 1, how much the net recalls well-understood linguistic relations when the data has been annotated with them. But often we don't have such gold-standard annotations at hand, and we may have to make do with noisy machine annotations. What we don't have a handle on is the significance of other kinds of middling-strength relations for which there doesn't appear to be a recognizable linguistic pattern. Some may be internal inferences needed to build up valid relations, others may be relics of imperfect learning, like our wisdom teeth or the tails that briefly appear in human embryos.

How does the model handle long-distance context across sentences? Technically, all input words are turned into a arrays of numbers, and then queries, keys, and values are computed using learned weights. The model compares each query with all keys to decide how much attention to pay to each word and combines the corresponding values accordingly. To prevent peeking at future words (important for language modeling), it uses a technique called *masking* to block access to later positions in the sequence. The details are spelled out in the next section.

2.5.4 Handling Richer Context*

> Now consider an even wider prior context. Among the keys in the prior context that it could consider is k_{10} from the Candidate sentence, derived from the word 'candidate'. The attention weight captures the relevance of 'candidate' to 'hunter'. The output a_2 in the Hunter sentence, using Eq. 2.13, incorporates the value vector $v_1 0$ from 'candidate' in the Candidate sentence, as well as other value vectors, each with a different weight, thus integrating information from the preceding sentence into the representation of 'hunter'. This presumably allows the model to interpret 'hunter' as targeting a candidate, leading to an inference eventually of his being an assassin.
>
> The output for all words i from 1 to L, the length of the input, can be computed as a vector of outputs for each a_i. In fact, the entire input $\langle x_1 \ldots x_L \rangle$ can be turned into a matrix X of dimension $L \times d_m$, where d_m is the dimension of the word embedding. We will call $L \times d_m$ the original input shape. (In our example, $\angle x_1 \ldots x_L \rangle$ may cover all the words

2.5 Self-attention

from the Candidate sentence through the Hunter sentence in sequence.) Given this matrix X, the query vector Q (across all the input) will be computed by multiplying X with W_Q (of dimension $d_m \times d_m$) to yield the combined query matrix Q (of the original input shape). Likewise, we will compute K and V, which will each also be of the original input shape. Finally, taking the softmax of the $Q \times K^T$ matrix (itself of dimension $L \times L$) and multiplying by V, of the original input shape, we get the output, also of the original input shape:

$$Q(X) = X \times W^Q$$
$$K(X) = X \times W^K$$
$$V(X) = X \times W^V$$
$$\text{Self-Attention}(Q(X), K(X), V(X)) =$$
$$\text{softmax}\left(\frac{Q(X) \times K^T(X)}{\sqrt{d_k}}\right) \times V(X)$$

(2.14)

Note that the $Q \times K^T$ calculation compares each query value to every key value, including those that follow the query. This is inappropriate in the setting of language modeling, where our model doesn't have access to the next word. To fix this, the elements in the upper-triangular portion of the matrix are zeroed out, eliminating any knowledge of words that follow in the sequence. This, as we have seen, is called *masking*.

2.5.5 Multi-head Attention

2.5.5.1 Introduction

Since there may be many different kinds of relationships among words, at different distances from each other, attention can be refined further by introducing multiple self-attention layers, called (somewhat misleadingly) *Multi-head attention*. The idea is that each attention head can focus on different parts of the input. Each head might also discover different kinds of relationships. For example, one head might pick up syntactic dependencies, another head semantic relationships, a third head some other association. Note that what the heads pick up is discovered by the net, not pre-specified, so the mileage of different heads could vary. Having multiple heads is also a way of avoiding putting all one's eggs in one basket (if you will pardon the egregious mixing of metaphors). If one head is misled by making errors, another may make up for it.

2.5.5.2 Details*

Each head will have its own weight matrices for query, key and values. Assuming L is the length of the input, in head_m, we have weight matrices Q_m, K_m, and V_m. However, since there are multiple heads, these matrices have to be reshaped to split the information across the heads. Let us examine this briefly.

Instead of the original input shape, they will be of shape $L \times h \times d_m$. The embedding size also has to change, so we divide it by h. These three attention matrices are thus of size $\frac{L \times h \times d_m}{h}$. The different head outputs (each of dimension $\frac{h \times L \times d_m}{h}$) are then *concatenated* together to yield a new matrix (of the original input shape). This new matrix is multiplied with a new output weight matrix W_O (of dimension $d_m \times d_m$). The output is a matrix A of the original input shape.

The computations for h different attention heads are shown in Eq. 2.15.

$$\begin{aligned} A &= \text{Multi-head Attention}(X) \\ &= W_O \times \bigoplus_{m=1}^{h} \text{Self-Attention}(Q_m(X), K_m(X), V_m(X)) \end{aligned} \quad (2.15)$$

2.5.6 Summary

Now that we are done with the major equations and arithmetic, it's time to pause to summarize what we have discovered so far about modeling context. Embeddings are discovered through self-supervised machine learning, mining a treasure-trove of distributional relationships lying hidden in training data. The embeddings result in words with similar context being placed near each other in vector space, as with 'aimed' and 'fired' in the Hunter sentence. Self-attention zooms in on key relationships, and thus we found from inspecting the attention weights that 'aimed' now has not only the original embedding information but information from embeddings of related words like 'hunter' and 'crouched'.

Opacity and Emergent Knowledge
It's possible to speculate that the emergent knowledge is something along the lines of hunting involving a crouching position and aiming being a key action in the process of firing a weapon used in hunting. The problem, as we have noted repeatedly, is that the weights in neural nets are entirely opaque, and so far reverse engineering doesn't make explicit what such knowledge is, at least not in any human-interpretable terms, unless we have agreed-upon annotations with that information. We can focus on what the net is attending to, but unless we have some sort of hybrid neurosymbolic architecture, it's going to be difficult to address this problem.

2.6 Transformers

A popular and widely-used neural architecture that was developed in conjunction with attention models is called the Transformer. It begins with a rich representation of input word meanings based on their embeddings. It then transforms the input representation as it

2.6 Transformers

advances through its many layers, at each layer adding deeper relations of meaning among words, based on distributional semantics.

These relations, in turn, are based on more advanced representations of a word's context, incorporating local as well as longer-distance dependencies, including those from previous sentences.

All of these capabilities rely on a healthy marriage of distributional semantics and scalable computing using deep nets.

Before continuing, it's worth noting that in all the discussion of word embeddings (which are of course numerical representations of meanings based on distributional semantics), we didn't once mention the position of each word in the original input. Because the order of words in a sentence matters, we also need numbers that record the position of each word in the original input. (Word order in a sentence matters a lot in most languages, including English, and it matters to at least some extent in even such flexible word-order languages like Hungarian, Turkish, and Warlpiri.) Thus, the original input consists of a word embedding along with a positional embedding. This input is fed to the Transformer.

2.6.1 Introduction

A *Transformer* is made up of stacks of components called *transformer blocks*. Figure 2.8 shows a single block of the pipeline. (As we shall see, GPT-3 has 96 of these blocks stacked together; more advanced architectures have many more.)

The first component is the *MultiHead Attention Layer*, which represents multi-headed attention. If there are six heads, each of the heads, with its own query, key and value weight data, will attend *in parallel* to the input. This results in increased scalability as well as richer modeling of context.

There is also a shortcut that appends the original input to the output of this layer. This input-to-output agglomeration is called a *Residual Connection*, which in general helps preserve input to earlier layers during processing by later layers. When the net is very deep with many layers, the residual connections also allow the net to focus on the differences from the previous layers (hence the name 'residuals'), and thus preserving input information.

The next component, *Layer Normalization*, normalizes the values of outputs so that they fall within a pre-defined range. Layer normalization contributes to faster performance, as well as better efficiencies during training.

The component after that is a *Feed Forward Layer*, a net with a set of 2-layer FF NNs, with one FF NN at each input position. Since the FF NNs are independent of each other, they can be computed *in parallel*. This makes them far more scalable to large datasets.

Another *Residual Connection* hops over this component, appending its input to its output.

Next comes yet another *Layer Normalization*, which generates the final output of the Transformer block.

Fig. 2.8 Transformer block architecture

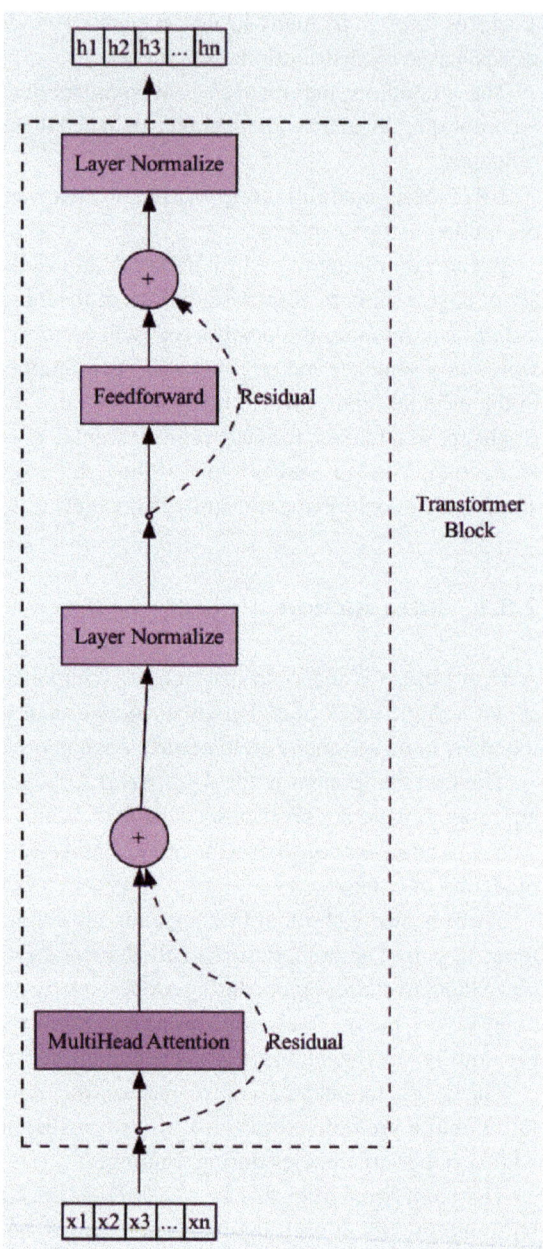

All this is just one Transformer block. A modern state-of-the-art LLM will stack together hundreds or more of these blocks, as we shall see in Sect. 2.6.3. This results in much deeper and more capable nets.

2.6.2 Architecture*

As mentioned earlier, *Transformers* are made up of stacks of *transformer blocks*, each of which is a multilayer network that maps sequences of input vectors $\langle x_1 \ldots x_L \rangle$, where each x_i is a vector of dimension d_m, to sequences of output vectors $\langle z_1 \ldots z_K \rangle$.

Input vectors x_i represent, as we mentioned earlier, the embedding of a particular subword. In practice, we sum the embedding for x_i with a position embedding, to represent information about input word positions and distances between words. Where do we get these positional embeddings? The simplest method, called absolute position, is to start with randomly initialized embeddings corresponding to each possible input position up to some maximum length. For example, given the embedding for the second word 'hunter' and the fifth word 'core' in the Hunter sentence, we'll add in embeddings for position 2 and 5 respectively. These position embeddings can then be learned during training.

As also mentioned earlier, transformer blocks are made by combining several layers: an attention layer, a normalization layer, a FF NN layer, and then another normalization layer, along with residual connections that hop across layers. Residual connections, mentioned earlier, also allow gradients to flow back through the network skipping intermediate layers, thereby lessening the vanishing gradient problem.

The *attention layer* uses Multi-head attention. (Recall that the attention layer generates output of the same dimension as the input. So we have its output as $A = \langle a_1 \ldots a_L \rangle$, where each a_i is a vector of dimension d_m.)

Layer normalization standardizes the values in the incoming vector A by each value's z-score, namely scaling each value by subtracting the mean from the value and dividing by the standard deviation. The output of layer normalization is an intermediate vector O of the original input shape.

The *FF NN* layer consists of a single FF NN with multiple hidden layers. The FF NN applies in parallel to each input token. The output B of the original input shape is then fed to another layer normalization layer, to generate the output Y of the original input shape.

Putting this all together, an input vector X of the original input shape is processed by the transformer block to yield the output Y of the original input shape:

$$\begin{aligned} A &= \text{Multi-Head Attention}(X) \\ O &= \text{Layer Norm}(X + A) \\ B &= \text{FF NN}(O) \\ Y &= \text{Layer Norm}(O + B) \end{aligned} \quad (2.16)$$

2.6.3 Language Modeling with Transformers

2.6.3.1 Introduction

Once we have run a sequence of words (or tokens) through a Transformer, how does it predict what comes next? We have already seen, in our Candidate sentence, how the prior context for our paradigm Hunter sentence can result in an inference that the hunter is an assassin. To make the continuation a little less predictable, let's now edit the paradigm Hunter sentence as in Example 2.7.

Example 2.7
A coward at heart, he crouched low, aimed and ....

We already saw in Chap. 1 that n-gram language models were inadequate for longer-distance dependencies. You do need rich representations of word meaning (from embeddings) as well as semantic relations found in the current prompt context. Some of these relations may come from the context of previous sentences, but here, one could expect that the information that the person is a coward comes from the current sentence. The attention mechanism can not only establish that 'he' is coreferential with 'coward', but can also influence 'aimed', which now has not only the original embedding information but information from embeddings of related words like 'coward' and 'crouched'.

Thus, on processing the last word 'and' in Example 2.7, that word now has information coming from several earlier places in the whole Coward sentence. This, one would expect, would help the Transformer make an inference as to what a coward with a weapon might do next, without having to have explicit knowledge of such things. Thus not only will 'fired' and 'squeezed' and 'shot' get a reasonable score, but so might 'fled' and 'hesitated', which are some of the things a coward might do in a corpus,

In the Transformer, all the words in the vocabulary will get scores, and these are converted by softmax into probabilities for each alternative, with the probabilities all adding to 1. To choose which word is next, the model might pick the top-ranked word based in the probability ranking as a default. To encourage more diverse or more creative choices, based on the user raising the Transformer's *temperature* parameter, which can look beyond top-ranked words for additional next word choices. Of course, if the temperature is set too high, the Transformer may make feverishly random predictions, which defeats the whole purpose.

The language model in the transformer is implemented by a Language Modeling Head, whose mathematical details are discussed next.

2.6.3.2 Language Modeling Head*

How do we use a Transformer for language modeling? Given the final hidden layer Z of the Transformer, we take that layer's state, namely the output z_n corresponding to the last word of the input. In our Hunter sentence, for example, that word was token x_n (where $n \leq L$), which was 'and'. Thus z_n has all the contextual information after processing x_n. The information from z_n is then used to produce a probability distribution of possible next words and generate one of them. It does this via a block called the *Language Modeling Head*.

2.6 Transformers

The first module of the Language Modeling Head is a *projection layer*, which takes z_n's vector of shape d_m and transforms it to to a vector that represents scores for the possible next word. The projection layer's output vector v_n will thus be of size d_{voc}, and its values are real numbers. The vector is computed by a linear combination of a weight matrix W of shape $d_m \times d_{voc}$ and a bias vector B of size d_{voc}, reflecting a bias over the entire vocabulary. The vector v_n is then turned into a probability over the vocabulary using softmax to yield an output vector Y_n:

$$v_n = z_n \times W + B$$
$$Y_n = \text{softmax}(v_n)$$
(2.17)

So, now at each time step, given all the possible next words from the vocabulary with their probabilities, the Language Modeling Head has to choose, or sample, one as the output word x_{n+1}. It usually does this by combining two factors. The first factor is k: the top k most probable words given the words it has generated so far, where k is a hyper-parameter that can be adjusted for the application. This is a form of greedy decoding, where the top k words are are renormalized to form a probability distribution, where all the probabilities add to 1, and then the most probable word is chosen from that distribution. However, with a low-entropy probability distribution, where the choices are more or less equally likely, the top-k method may end up selecting a low-probability word, and with a high-entropy one, may not be diverse enough in its selection. As Holtzman et al. (2020) discovered, the text produced by this method can be incoherent and unrelated to the context—and the reason for this is the *unreliable tail* of the probability distribution, consisting of "tens of thousands of candidate tokens with relatively low probability that are over-represented in the aggregate." Their alternative to top k sampling is *nucleus sampling*, where instead of choosing the top k most probable words (which could include low-probable words), just enough words are chosen so that their combined probability exceeds a (say 90%) threshold. We shall have more to say about this when it comes to story generation, in Chap. 6.

The second factor is T, the *temperature*, a hyper-parameter that introduces randomness into the predictions, aimed at generating more diverse or creative outputs. To use temperature T, the values in vector v_n will be divided by T. A lower temperature (e.g., $T = 0.7$) will stick close to the probability, whereas a higher temperature (e.g., $T = 1.5$) will flatten the distribution to favor less likely next words. However, a high temperature can lead to repetitive texts when the probability distribution is multimodal (i.e., has multiple peaks).

> **Prediction controls via hyper-parameters**
> Unfortunately, choosing values for these hyper-parameters can involve a lot of unsatisfactorily ad hoc tuning (or knob twiddling) and even then results may not be satisfactory.

> As Basu et al. (2021) have discussed, low values of k result in repetition, and large values result in incoherent text, even with nucleus sampling. They find that repetitions in generated texts are correlated (unsurprisingly) with dips in *surprisal* values, whereas incoherence is correlated with large and increasing surprisal values across the sequence. The authors' neural approach allows specification of a target surprisal, and a surprisal of 0.3 tends in their human evaluations on their datasets to result in high-quality texts.

2.6.3.3 Further Discussion

While we are on the subject of predicting the next word, attention has also to be paid to not generating biased or discriminatory text. A perfectly reasonable word in isolation can, after all, given a perfectly reasonable prior context, produce something horrible. "Playing the game" may be fine, but "Playing the victim" may not.

Once we have generated a safe next word, it is appended to the Transformer input, and processed accordingly. This it becomes part of the context for the generation of the word after that. Thus, the Transformer generates an output word, or token, at a time, conditioning each next-token prediction on the previously generated tokens, a process that we mentioned earlier, called *autoregression*.

Let us examine the probabilities of next words given the partial sentence in Example 2.3, namely "A hunter to the core, he crouched low, aimed and ...". Here I used the model 'gpt-neo-2.7B'. Note that altering various hyper-parameters could change some of the output probabilities, so their values should not be taken as definitive. The results are shown in Table 2.1.

If we ask the net to keep going, it suggests the follow-up words "fired his rifle with precision at the elusive target". A little cliched, but nevertheless it conveys the idea of what is going on. If we insert a delimiter like *tl;dr* (too long, don't read) after a much longer input text, to indicate a summary is desired, the machine can go on to produce a summary which is the basis for many approaches to summarization today—though LLMs may have

Table 2.1 LLM Next Word Probabilities for sentence in Example 2.3

Word	Probability
'fired'	0.7574
'pulled'	0.0839
'squeezed'	0.0828
'shot'	0.0567
'let'	0.0192

difficulty, unlike some of the pre-LLM summarization algorithms, in achieving the desired target length.

Putting the different Transformer steps together, it is clear from our probes that the embeddings for 'aimed' and 'fired' had an affinity for each other. The self-attention revealed that 'aimed' is in a relationship with 'crouched', which in turn is in a (potential subject-verb) relationship with 'he', with both 'aimed' and 'crouched' being also in a (potential subject-verb) relationship with 'hunter'. With all these relationships firing on all cylinders, the language modeling head is able to put forward 'fire' as its strongest guess, with the other words consuming much smaller parts of the probability distribution.

2.6.4 Sample Specs for a Transformer*

Let's take a peek at the specs of GPT-3, the 175 billion parameter version. The input fed to the Transformer is batched, so we have a batch size $= b$ set of inputs being fed in. The original input shape is therefore $b \times L \times d_m$. The vocabulary size d_{voc} is 50257 tokens. The embedding is of dimension d_m is 12288. The maximum input length L is 2048 tokens. So the original input shape is $b \times 2048 \times 12288$. The tokens are byte-pair encoded, which means that roughly 4 successive characters in the input is treated as an input token, so that each token is almost three-fourths of a word. That means that up to 1536 input words can be accepted at a time.

GPT3-175B has 96 transformer blocks. Let's zoom into a single block. The query, key and value matrices Q, K, and V are of dimension $\frac{b \times L \times h \times d_m}{h}$, which is $b \times 2048 \times 96 \times 128$. The different head outputs (each of dimension $96 \times 2048 \times 128$) are then concatenated together (using the $\oplus$ operation) to yield a new matrix (of the original input shape). This new matrix is multiplied with a new output weight matrix W_O (of dimension 12288×12288). The output is a matrix A of the original input shape, as is the shape of O, B, and the transformer block output Y in Eq. 2.16.

2.6.5 Bidirectional Encoders

We now briefly overview a class of Transformers called *Bidirectional Encoders* like BERT (Devlin et al. 2019). They use all of the input to make predictions, not just the preceding input, so a prediction about a word uses both left- and right context. The final layer attention weights, which use all of the input, are often used as contextualized word embeddings for downstream tasks. This is contrasted with the lower-triangular attention weights we saw in Fig. 2.6.

Using bidirectionality, these nets are trained for masked language modeling, which corresponds to a Cloze task of filling in blanks (represented by a 'MASK' token, or some other random word), or next sentence prediction, which involves predicting whether a pair of sentences are related. The latter task can frame many applications, including detecting whether one sentence is a paraphrase of another, and determining logical or other relations between sentences, or other chunks of text. Google Search makes good use of BERT (Devlin et al. 2019) to analyze words in queries based on context on both sides to get a better grasp on what the user is searching for. However, BERT can't really generate text, as text generation can't be framed as a Cloze task unless one masks out every single word to be predicted next.

2.7 Fine-Tuning and RAG

2.7.1 Fine-Tuning

The goal of transfer learning is to take knowledge learned in one domain and use that as a starting point for learning new knowledge from another domain, on a possibly different task. This only works well when the task is similar and the distributions in each dataset are similar. Fine-tuning is a restricted form of transfer learning that involves training the model on more specialized data for a given domain, but for a similar task. It doesn't involve re-training, but additional training.

For example, one might take a LLM trained for summarization on the Web in the large and train it for summarization of internal company data, such as customer feedback message data. Given the computational costs of training a Transformer, developers often choose only some parameters to modify and *freeze* the rest, reducing the number of parameters drastically. This has the added advantage of the new, extended model not catastrophically forgetting most of the weighting of the parameters it learned from the original data. Comparing the source and target dataset distributions can help in that process.

One popular tool in the fine-tuning toolbox is instruction tuning, which can supplement or replace training on the new domain, depending on the use case. This method (Wei and Maarten 2022) uses few-shot prompt engineering techniques (explained in Sect. 2.8), providing an instruction and a few input-output pairs. For example, consider a particular version of a BERT Foundation LLM, which has already been trained for sentiment analysis. We could give the tool the instruction, say "Analyze the sentiment of the following passage", and provide several example of narrative passages and the desired labels. Other fine-tuning tools include reinforcement learning (RL) from supervised human feedback (or unsupervised—and noisy—AI feedback). In RL, the agent carries out an action which is then given a reward, and the agent learns a policy that maximizes the expected cumulative reward. Since it involves an unsupervised search, it usually requires a lot of data to train an effective policy.

Such RL methods are used by most advanced chatbots including ChatGPT to enforce cooperative behavior. ChatGPT uses the neural net to implement the RL. Recently, DeepSeek (DeepSeek 2025) used a rule-based RL method in its foundation model, where a group of outputs from the model is sampled and compared with each other, and given a reward based on a guess of how correct it is—relatively speaking. When the problem is a math problem, the correctness of the output can be based on calculation. For computer code generation problems, correctness can be checked by code compilation; for other problems they use heuristics (such as conciseness, relevance, etc.) The pitfall of needing a lot of data to train the RL policy is avoided by use of relative comparison against a few examples and rule-based rewards using a non-neural RL implementation.

2.7.2 RAG

Retrieval Augmented Generation (RAG), from Lewis et al. (2020), is a way of augmenting the performance of LLMs with external knowledge, whether in the form of unstructured documents or knowledge graphs. This allows the LLM to use more up-to-date, factual and domain-specific information. This can also help mitigate hallucinations.

In RAG, the user's prompt, treated as a query, uses a search engine to retrieve documents from a specified collection. The retrieved results are then used as additional context to extend the user's prompt (called input-layer integration). The retrieval and generation components are usually jointly trained, allowing the generation to learn patterns of relevance from the retrieved information. In case the retrieved information is too long, it can be filtered, summarized, chunked into manageable segments, or split into a sequence of prompts by the LLM. A closer integration can be achieved in principle through intermediate-layer integration. A considerable amount of research has looked into different integration architectures. For example, the Borgeaud et al. (2022) approach interleaves into the transformer blocks separate retrieval blocks into which word embeddings from the retrieved passages are fed, and new cross-attention layers are used to attend to the retrieved information at each step. The above RAG approaches can be helpful to some extent in reducing hallucinations.

2.8 Prompt Engineering

Understanding the user's intent in a way that the system can respond effectively is naturally very important. A prompt is basically additional context prepended to the prior sentence context for predicting the next word. Hopefully, the right additional context can provide more useful information for selecting the appropriate next word choices. Unfortunately, there is no science to the crafting of prompts. At best, we have only some suggestions for best practices for different domains. Being as precise and unambiguous and specific as possible can of course help.

As we shall see in Chap. 6, automatically generating prompts for story generation as a result of high-level planning can be fruitful. In this section, however, we focus on hand-generated prompts.

In-context learning is an unsupervised method where the prompt to the machine includes a series of input-output pairs, without updating the model's weights or other parameters.

This is also called few-shot prompting, not to be confused with the fine-tuning method of few-shot learning which is supervised.

CoT Prompting

One variety of few-shot prompting is Chain-of-Thought (CoT) prompting (Wei and Xuezhi 2022), where a prompt is first decomposed by the user into logical steps, with each step being given to the machine before the next.

CoT prompting has some successes in areas like arithmetic (at which LLMs are, not surprisingly, rather bad, since guessing what comes next is hardly a good strategy for arithmetic). It has also been used in commonsense reasoning. For example, from Wei et al. (2022), in Example 2.8, the system gives a wrong answer to second problem, a simple word problem in arithmetic.

Example 2.8

> **User Prompt**: Q1: Roger has 5 tennis balls. He buys 2 more cans of tennis balls. Each can has 3 tennis balls. How many tennis balls does he have now?
> QA1: The answer is 11.
> Q2: The cafeteria had 23 apples. If they used 20 to make lunch and bought 6 more, how many apples do they have?
> **Machine**: MA1: The answer is 27.

Example 2.9, shows how a CoT prompt—where reasoning is explicated in the prompt—results in the correct answer.

Example 2.9

> **User Prompt**: Q1
> QA2: Roger started with 5 balls. 2 cans of 3 tennis balls each is 6 tennis balls. $5 + 6 = 11$. The answer is 11.
> Q2:
> **Machine**: MA2: The cafeteria had 23 apples originally. They used 20 to make lunch. So they had $23 - 20 = 3$. They bought 6 more apples, so they have $3 + 6 = 9$. The answer is 9.

The fact that the CoT reasoning in the prompt gets emulated by the machine is, some enthusiasts would say, an instance of 'emergent thought' or 'thinking aloud'. However, while CoT can result in improved performance, there is the danger of overfitting, so generalization of the learned CoT reasoning across domains needs to be demonstrated. It is also hard to assess whether genuine reasoning is occurring, and there is inherent subjectivity in the assessment.

For analyzing narratives, being as precise as possible in the instructions, and iteratively refining the instructions, seems to work well as a CoT method. For generating narratives, as we shall also see, CoT can be helpful.

> **Why does in-context learning work?***
> An experiment by Chan et al. (2022) used the Omniglot image classification dataset, consisting of 1623 different character classes from various international alphabets, with each class containing 20 handwritten examples. The few-shot challenge was to classify an example of a character class unseen in the training data, based only on a few examples of that class and some alternate classes. The authors discovered that in-context learning can emerge in a Transformer only when the original training data includes both a large enough set of rarely occurring classes and the *burstiness* that characterizes natural language distributions in the large. Burstiness is the tendency for words not to be distributed uniformly over time, but to appear clumped together in bursts. It implies non-stationarity, meaning that tokens in one position can be dependent on tokens at arbitrary distances away in time. The bottom line is that the distributional properties of natural language have to be adequately mirrored in the training data for in-context learning to emerge, suggesting the possibility of studying the data distributions further to see what aspect of them can improve learning.

2.9 Assessment and Conclusion

This chapter began by introducing a neural model called a Simple Linear Perceptron, describing how to train it using Stochastic Gradient Descent (SGD), and pointing to its weaknesses on trivial problems like XOR inference. This led to a discussion of the more powerful Feed Forward Neural Net, illustrating its architecture and training using SGD. We then described several methods for calculating word embeddings based on distributional relationships lurking in data, which result in words with similar context being placed near each other in vector space. We discussed self-supervised machine learning methods used to build them. Next, we turned to richer models of sentential context using Self-Attention, progressing to handling wider context and using multiple attention heads to allow for different relationships among words, including between words that are further apart. Finally we introduced Transformers and how to use the architecture to do language modeling, including bidirectional encoders that look at both left and right context. The rest of the chapter dealt with use of Transformers as LLMs, including customizing it for different applications and domains, namely augmenting its inferences via additional knowledge sources using Retrieval-Augmented Generation and fine-tuning with more specialized training data. These two methods are not foolproof, and require quite a bit of experimentation to be successful. Finally we turned to prompts and how to design them, including using Chain-of-Thought prompting for narrative analysis.

The Transformer's huge scale and the power of Self-attention models have contributed to a startling breakthrough, allowing them to outperform other models on a variety of benchmark tasks. As Kaplan et al. (2020) has shown in their paper on scaling laws, the accuracy (measured in terms of decreased cross-entropy loss) improves rather dramatically in a power law relationship as one increases model size, dataset size, and the amount of compute used for training.

Nevertheless, Transformers have several major disadvantages:

1. **Cost**:
   Training such models from scratch is very compute-intensive and is of quadratic time complexity, and it also requires quadratic memory. Inference is also relatively slow, because the Self-attention mechanism is of quadratic time complexity in its comparisons, and autoregressive generation of one token at each step requires recomputing self-attention for all tokens generated so far. The enormous cost also puts training out of reach of most research groups.
2. **Size Bias**: Transformers are data-hungry, and don't work well when trained on smaller datasets. However, bigger is not necessarily better. For example, Lialin et al. (2022) tested 29 different Transformer models and found that the number of parameters, dataset size, and other features do not correlate with model performance in certain tasks. Not every task requires a LLM. Smaller language models (with only millions of parameters), which do not require high-end GPUs for training and that can carry out fast inference, have their niche in real-time applications.
3. **Test set availability**: Since these systems require training on huge quantities of data, of the sort available off the Web, it's hard to really verify that apparently held-out test sets haven't in some form been seen before. Likewise those vast datasets almost certainly include annotated data. Accordingly, the era of tightly-controlled experiments one could run by training from scratch on small-scale datasets with held-out test sets appears to be gone. Nevertheless, the performance on benchmarks as well as examples throughout this book suggests that these so-called 'unsupervised' neural methods are doing well anyway, so that it may not matter that much.
4. **Context limitations**: The Self-attention mechanism weakens as the contexts get longer, and with longer-form narrative understanding or generation, this poses a substantial challenge.

Alternatives to Transformers are gaining ground, including Selective State Spaces (Gu and Tri 2024), which are far more efficient through not paying the quadratic time cost of attention. They nevertheless handle long-distance dependencies efficiently. It seems clear that we are in a stage where there is a plethora of different approaches emerging, but the core ideas of neural computing, and exploiting context and distributional semantics are here to stay. That is good news for the future of narrative in the context of Gen AI, provided we can address problems of efficiency and scaling, and, in the longer term, conquer some of the evaluation challenges as well as the limitations of LLMs and AI that the reader might recall from Chap. 1 (Sect. 1.6).

References

Basu, S., G. S. Ramachandran, N. S. Keskar, and L. R. Varshney (2021). "Mirostat: A neural text decoding algorithm that directly controls perplexity". In: *International Conference on Learning Representations*. URL: https://openreview.net/forum?id=W1G1JZEIy5%5C.

Borgeaud, Sebastian et al. (2022). "Improving Language Models by Retrieving from Trillions of Tokens". In: *Proceedings of the 39th International Conference on Machine Learning*, Baltimore, Maryland. URL: https://proceedings.mlr.press/v162/borgeaud22a/borgeaud22a.pdf.

Chan, Stephanie C.Y. et al. (2022). "Data Distributional Properties Drive Emergent In-Context Learning in Transformers". In: *Proceedings of NeurIPS*. URL: https://openreview.net/forum?id=lHj-q9BSRjF.

DeepSeek-AI et al. (2025). *DeepSeek-R1: Incentivizing reasoning capability in LLMs via reinforcement learning*. URL: https://arxiv.org/abs/2501.12948.

Devlin, J., M.-W. Chang, K. Lee, and K. Toutanova (2019). "BERT: Pre-training of deep bidirectional transformers for language understanding". In: *Proceedings of NAACL HLT*. URL: https://www.aclweb.org/anthology/N19-1423.

Gu, Albert and Tri Dao (2024). "Mamba: Linear-Time Sequence Modeling with Selective State Spaces". In: *First Conference on Language Modeling*. URL: https://openreview.net/forum?id=tEYskw1VY2.

Holtzman, A., J. Buys, M. Forbes, and Y. Choi (2020). "The Curious Case of Neural Text Degeneration". In: *Proceedings of International Conference on Learning Representations, 2020*. URL: http://arxiv.org/abs/1904.09751.

Jurafsky, Dan and James H. Martin (2024). *Speech and Language Processing*. URL: https://web.stanford.edu/~jurafsky/slp3/.

Kaplan, J. et al. (2020). *Scaling laws for neural language models*. URL: https://browse.arxiv.org/pdf/2001.08361.

Lewis, Patrick, Ethan Perez, Aleksandra Piktus, Fabio Petroni, and Vladimir Karpukhin (2020). "Retrieval-Augmented Generation for Knowledge-Intensive NLP Tasks". In: *Proceedings of NeurIPS*. URL: https://arxiv.org/abs/2005.11401.

Lialin, Vladislav, Kevin Zhao, Namrata Shivagunde, and Anna Rumshisky (2022). "Life after BERT: What do Other Muppets Understand about Language?" In: *Proceedings of the 60th Annual Meeting of the Association for Computational Linguistics (ACL)*, URL: https://aclanthology.org/2022.acl-long.227.pdf.

Mikolov, T., I. Sutskever, K. Chen, G. S. Corrado, and J. Dean (2013). "Distributed representations of words and phrases and their compositionality". In: *Proceedings of NeurIPS*. URL: https://proceedings.neurips.cc/paper%5C_files/paper/2013/file/9aa42b31882ec039965f3c4923ce901b-Paper.pdf.

Pennington, J., R. Socher, and C. D. Manning (2014). "GloVe: Global vectors for word representation". In: *Proceedings of EMNLP*. URL: https://doi.org/10.3115/v1/D14-1162.

Rosenblatt, F. (1958). "The perceptron: A probabilistic model for information storage and organization in the brain". In: *Psychological Review* 65.6, pp. 386–408.

Vaswani, A. et al. (2017). "Attention is all you need". In: *Proceedings of NeurIPS*. URL: https://proceedings.neurips.cc/paper%5C_files/paper/2017/file/3f5ee243547dee91fbd053c1c4a845aa-Paper.pdf.

Wei, Jason, Maarten Bosma, et al. (2022). *Finetuned Language Models Are Zero-Shot Learners*. URL: https://arxiv.org/pdf/2109.01652.

Wei, Jason, Xuezhi Wang, et al. (2022). "Chain-of-thought prompting elicits reasoning in large language models". In: *Proceedings of the 36th International Conference on Neural Information Processing Systems (NeurIPS 2022)*. New Orleans, LA: Curran Associates Inc. URL: https://arxiv.org/abs/2201.11903.

Introduction to Narratology

3

> *She was merely a narratologist, a being of secondary order, whose days were spent hunched in great libraries scrying, interpreting, decoding the fairy-tales of childhood and ...the impeded couplings of doctors and nurses, dukes and poor maidens, horsewomen and musicians.*
>
> —A. S. Byatt (1994)

3.1 Introduction

Narratology is a theory of narrative structure, derived from literary criticism. It is "a humanities discipline dedicated to the study of the logic, principles, and practices of narrative representation", as stated in Meister (2014).

This chapter covers basic concepts from the field of Narratology. It begins by introducing the field itself, and then goes on to discuss narrator identity, distance, perspective and focalization. This leads to exploring different narrative levels, including embedded narratives, narrative threads, and subordinated discourse. The scope of different narrative elements brings up issues of which character is accessible to others at what level, which involves a further dip into the theory of meaning. Next, I introduce the concept of the audience, different kinds of readers, and audience response. The chapter is rounded up with the introduction of NarrativeML fragment that captures all the narratological distinctions so far. The framework is elaborated in subsequent chapters. Throughout, use is made of Gen AI models to see how they perform in narrative analysis and generation, focusing on these different facets of narrative and violations thereof.

The material in this chapter is presented from the ground up, without any requirement of prior exposure to the edifices of literary theory. This is likely to provide fertile ground for humanities readers, but it is likely that scientists will also benefit. For example, since some of the NLP and AI work on narrative seems to lack an awareness of narratology and its various distinctions, improving that awareness can lead to more insightful benchmarks and development methods for Gen AI systems.

Although its antecedents lie in classical theories of aesthetics, narratology derives more directly from the intrinsically language- and linguistics-centered approaches to literature fostered by the Russian formalists in the early years of the 20th century and later through European structuralism, among whom the most influential names are Russian and French scholars like Bakhtin, Barthes, Bremond, Genette, Propp, Shklovsky, and Todorov (the last of whom, in 1969, introduced the term *narratology*). In addition to studying traditional topics such as plot and character, these scholars also provided compelling accounts of the way time and point-of-view are expressed in narrative. These approaches make narratology especially interesting to computational and corpus linguists.

The 1980s onward has seen the growth in the humanities of *post-classical* narratology, which retains an interest in language but which is focused around two active subfields:

1. cognitive narratology, i.e., "the story-producing activities of tellers, the processes by means of which interpreters make sense of the narrative worlds (or *storyworlds*) evoked by narrative representations or artifacts, and the cognitive states and dispositions of characters in those storyworlds" (Herman 2014);
2. contextualist narratology, which, inspired by European 'theories' of deconstruction, relates "the phenomena encountered in narrative to specific cultural, historical, thematic, and ideological contexts" (Meister 2014). While no doubt interesting, contextualist narratology is less relevant to computation, in part because computation has not advanced enough to deal adequately with such topics.

The narratological developments in the humanities have been paralleled in the last half-century by the growth of computational approaches to narrative. This has resulted in systems for generating and understanding stories. More recently, the demands of interactive entertainment, and interest in the creation of engaging narratives with life-like characters, have given rise to interactive fiction, drama, and advanced computer game environments. These endeavors have created computational instantiations of narrative constructs, formulating (sometimes implicitly) theoretical and empirical approaches to narrative. **Computational narratology** examines narratology from "the point of view of computation and information processing" focusing on, from Mani (2014):

> the algorithmic processes involved in creating and interpreting narratives, modeling narrative structure in terms of formal, computable representations. Its scope includes the approaches to storytelling in artificial intelligence (AI) systems and computer (and video) games, the automatic interpretation and generation of stories, and the exploration and testing of literary hypotheses through mining of narrative structure from corpora.

Some literary scholars (Scholes 1996) have characterized the earliest non-dramatic narratives as being epics, which involve oral retellings of a traditional story. Over time, epic gives rise to two varieties of narrative:

1. Empirical narratives that involve allegiance to reality, emphasizing truth (i.e., what really happened, and what most likely caused something). These include histories, biographies, autobiographies, and character sketches (such as those by the Greek Theophrastus in the 4th century BCE);
2. Fictional narratives that involve allegiance to an ideal, emphasizing aesthetic notions such as beauty and moral ones such as goodness. These include romances and didactic works such as satires, fables, anecdotes, allegories, and hagiographies (e.g., the 4th century BCE partially fictionalized biography of the Persian Emperor Cyrus, the Greek *Cyropedia*).

These two strains come together, (Scholes 1996) argue, in early Latin novels, such as *The Golden Ass* (2nd Century), and culminate eventually in the modern novel with all its different experimental forms. Although such a historical account might have to be revised substantially to take into account the much vaster literatures of non-Western cultures, not to mention the need to factor in the evolution of poetry, drama, the short story, and the impact of new media, it does give an idea of the breadth of various genres of literary narrative that narratology has to contend with.

3.2 Narrator Characteristics

3.2.1 Narrator Identity

The narrator is by definition present in all narratives, fictional or non-fictional; her identity is the answer to the question *"Who speaks?"*. If my friend is telling me (the audience) a story in my presence, or over the phone, or by email, she is clearly the narrator. The narrative in question could well be a joke or folk-tale attributed to someone else; in such a case, the narrator is not the **author**—the latter is the creator of the story.

In the case of a *first-person* narrative that is non-fictional, the narrator is identified with the author, whereas in the case of a first-person fictional (or poetic) narrative, the narrator need not be the author. Thus, when Maxim Gorky begins the second volume of his autobiography, called *My Apprenticeship*, with the words "And so my apprenticeship began", it is fair to treat "my apprenticeship" as coreferential with Gorky's apprenticeship. However, when W. G. Sebald begins his novel *Austerlitz* with "In the second half of the 1960s I traveled repeatedly from England to Belgium", the "I" is coreferential with the fictional Jacques Austerlitz, the narrator of the novel. In the case of a first-person narrative such as Joseph Conrad's *Heart of Darkness*, the overall first-person plural narrator is not identified, but the bulk of the narrative is told in the first-person by a character in the story called Marlow, who is an embedded narrator. Narratives such as the Sanskrit *Panchatantra* and *Mahabharata*, and the Arabic *The Thousand and One Nights* are filled with characters who tell other stories, allowing for multiple levels of discourse nesting. I will say more about such embedded narratives in due course. For the time being, it is worth bringing in a distinction here, from Genette (1980), between a **homodiegetic** narrator like Marlow, who is part of the story he tells, and a **heterodiegetic** narrator, who isn't, as in the case of the unnamed narrator of many third-person narratives.

In some narratives, the overall narrator has only a weak or covert presence. For example, Ernest Hemingway's short story *The Killers* is mainly dialogue in direct speech, and Virginia Woolf's novel *The Waves* consists mostly of direct speech monologues by different characters. Choderlos de Laclos' *Les Liaisons Dangereuses* is a collection of letters, i.e., an epistolatory novel, preceded by a spurious 'Publisher's Note' and an 'Editor's Preface'. In other narratives, the narrator's presence may be more overt: he may comment on the characters, or the veracity of his story, or ask the reader to skip certain parts, or draw attention to himself and his narrative effort in various ways. Sometimes a narrator may repeatedly interrupt, digress, and even question the whole narrating endeavor, as in the case of the over-intrusive narrator of Laurence Sterne's *The Life and Opinions of Tristram Shandy, Gentleman*. All these degrees of exposure can often help the author in her narratological goals, and in Sterne's case, the digressions are central to the whole satirical nature of the work.

In narrative understanding systems that map from text input to internal representations such as a fabula, the narrator is defined as above, namely as the answer to the question "Who speaks?". In narrative generation systems that take in a fabula or other representation and output coherent text, with or without audience interaction, the narrator is the system, or a specific component of it, such as the *Teller* component of Montfort (2011).

LLMs on Narrator Identity
What happens when we ask GPT-4o about narrator identity? The answer is in Example 3.1 (my underlining is not shown to the LLM). Here the LLM's task can be viewed as replacing some of the noun phrases with blanks, and choosing in each case between three possible antecedents: the narrator, M. Jumbert, and John Jay.

Example 3.1

> **User Prompt**: Foreword to the Text: the writer of this strange text, M. Jumbert, died of coronary thrombosis in 1952 ... John Jay, Ph.D. 1956. Text: I was born in 1910, in Paris. My father was a gentle, easy-going person ...
> Who is the speaker in the Text?
> **Machine**: The speaker in the text is the narrator, M. Jumbert, who is recounting their own life. The foreword, written by John Jay, Ph.D., introduces the text and notes the author's death in 1952, but the main text is from the perspective of Jumbert, who starts by saying "I was born in 1910, in Paris." Thus, Jumbert is the speaker, and the text appears to be a personal account or memoir.

My experiments show that LLMs are able to establish narrator identity in the presence of multiple distractors, revealing that it is a relatively easy anaphoric and coreference resolution problem for them.

3.2.2 Narrative Distance

A narrator can vary the **narrative distance**—another term from Genette (1980)—between herself and the characters, using different forms of speech to invoke the 'speech' of other narrators. This notion goes back to Plato, who distinguished between *simple narration* or *diegesis* on one hand and *imitation* on the other. Imitation is more traditionally called *mimesis*, from Plato (1871). Diegesis corresponds to what is informally called *telling*, whereas mimesis corresponds to *showing*.[1] In diegesis, the poet is the only speaker and his expression relies on indirect speech, as in Example 3.2, which is Plato's (prose) paraphrase of a fragment from Homer's *Iliad*, from Plato (1871) (III):

Example 3.2
The priest came and prayed the gods on behalf of the Greeks that they might capture Troy and return safely home, but begged that they would give him back his daughter, and take the ransom which he brought, and respect the God. Thus he spoke, and the other Greeks revered the priest and assented. But Agamemnon was wroth, ...

The mimetic version involves direct speech, where the narrator 'takes on' the role of a character in the story, imitating him. As shown in Example 3.3 in the corresponding passage from the *Iliad* (in the poetic translation by Fitzgerald 1975), here the narrator first imitates Khryses, and then the soldiers (ibid., p. 12):

Example 3.3

This priest, Khryses, had come down to the ships
with gifts, no end of ransom for his daughter;
on a golden staff he carried the god's white bands
and sued for grace from the men of all Akhaia,
the two Atredai most of all:
"O captains
Menelaos and Agamemnon, and you other
Akhaians under arms!
The gods who hold Olympos, may they grant you
plunder of Priam's town and a fair wind home,
but let me have my daughter back for ransom
as you revere Apollo, son of Zeus!"
Then all the soldiers murmured their assent:
"Behave well to the priest. And take the ransom!"
But Agamemnon would not. It went against his desire,
...

---

[1] Creative writing instructors on American campuses often exhort students to show, not tell, apparently unaware of the virtues of each mode and mixes thereof.

As can be seen, mimesis can often occur in the company of diegesis, as direct and indirect speech can be mixed. Literature in the form of stories and novels have greatly enriched the techniques by which the narrator can relate to the characters. Based on the approach of Genette (1980) and his revised account in Genette (1988)), we can distinguish three points on the distance continuum:

1. **Narrated Speech** (*Discours Narrativisé*): Here the narrator describes a character's speech or thoughts without directly quoting their words. It can be used by a first person narrator or a third person narrator. Examples: *I informed my colleagues of my decision to resign.* or *She decided to resign before they could expel her for speaking out.*
2. **Transposed Speech**: There are two subtypes:

    a. **Indirect Style** (*Discours Indirect*): The character's words are transposed into the narrator's discourse using subordinate clauses. Example: *She said that she was resigning the next day.*
    b. **Free Indirect Style** (*Discours Indirect Libre*): The narrator's discourse and the character's voice are *blended* together, omitting reporting clauses like "She thought" or "She said". It is used by a third person narrator who expresses a particular character's point of view. It is perhaps the most literary and flexible of forms of omniscient narration (see the next Sect. 3.2.2) and manages to reflect both the narrator's and the character's perspective. One of the most striking uses of it is in Virginia Woolf's *Mrs. Dalloway*, where the older Mrs. Dalloway is recalling her younger self:

Example 3.4
How fresh, how calm, stiller than this of course, the air was in the early morning; like the flap of a wave; the kiss of a wave; chill and sharp and yet (for a girl of eighteen as she then was) solemn, feeling as she did, standing there at the open window, ...standing and looking until Peter Walsh said, "Musing among the vegetables?"—was that it?—"I prefer men to cauliflowers"—was that it?.

3. What we might call **Quotation**; here there are three varieties:

    a. **Reported Speech**: Here the narrator "pretends to give the floor to his character" (Genette 1980). The character's words are expressed without subordinated clauses: *I thought/said to my mother: it is absolutely necessary that I marry Jake.*
    b. **Direct Speech**: (*Discours Direct*): The character's words are enclosed in quotation marks. Example: *I thought/said to my mother: "it is absolutely necessary that I marry Jake."*
    c. **Immediate Speech** (also called **free direct speech** or **interior monologue**): The narrator takes on the role of the character, expressing his inner thoughts and speech. This is seen in the great soliloquies of drama and fiction.

Narrated Speech and Transposed Speech narrate events or opinions rather than quote actual words verbatim, which differentiates them from Quotation.

These distinctions are easily discerned by LLMs, as they refer to notions in the already-seen literature.

3.2.3 Narrator Perspective: Focalization

A narrator's perspective, or point-of-view, is also a crucial aspect of narrative. It is the answer to the question *"Who sees?"*. A narrator may take the point-of-view of a particular character for part or all of the narrative, or she may be omniscient, 'seeing' into the mind of one or more characters. **Focalization** is the term used by Genette (1980) to represent perspective, based on the distinctions offered by Todorov (1981)and others.

Genette proposes three mutually exclusive and exhaustive cases, based on disparities in knowledge between narrator and characters:

1. In a **non-focalized** (or **zero-focalized**) narrative, the narrator is *omniscient*, knowing more than any of the characters, and saying more than any of the characters. The narrator has maximal access to the mind and spatial and temporal perspective of any of the characters; he can also foresee what will happen to them. Examples are ubiquitous, including the novels of Jane Austen.
2. In an **internally-focalized** narrative, the narrator sticks to what a given character knows. The idea here is that the narrator can't see more than the character does, while having complete access to the mental states of the character. In this case, the narrator expresses the point-of-view of a given character (called the **focalizer**), either a single one for the entire narrative, or varying within the narrative between one character and another, as in the case of Hemingway's *The Short Happy Life of Francis Macomber*.
3. In an **externally-focalized** narrative, "the narrator says less than the character knows" (Genette 1980, p. 189), a characteristic of 'behavioristic' narrative—such as Hemingway's *The Killers*. Here the narrator does not see into the characters' minds, or describe matters from their viewpoint; he pretends to be a non-judgmental witness.

In an internally-focalized narrative, the focalizer may provide less information about a character than is necessary; this is called **paralipsis**.[2] In Agatha Christie's *The Murder of Roger Ackroyd*, narrated by the character James Sheppard, the detective Poirot reconstructs the sequence of events involving the murder, revealing in the book's penultimate chapter, in conversation with Sheppard himself, why Sheppard is the killer. The last chapter of Sheppard's narrative is both a confession and a suicide note, indicating that he had hoped that his book would have revealed Poirot's failure in solving the case, rather than his success.

---

[2] Genette extends the use of the term in classical rhetoric, where *paralipsis* means "the device of giving emphasis by professing to say little or nothing of a subject, as in 'not to mention their unpaid debts of several millions'." (OED).

A focalizer may also provide too much information, which Genette calls **paralepsis**. The examples he cites, from Henry James' *What Maisie Knew* and Herman Melville's *Moby Dick*, are however, unconvincing, and literary scholars have since pointed out that most of those instances are simply examples of unreliable narration, a term first discussed by Booth (1983).

> **Digression: Dramatic Irony**
>
> In *dramatic irony*, the narrator reveals things about the characters or events in the story that the characters aren't aware of. This disparity between what the audience knows and what a character knows creates suspense and anticipation among the audience (and is a standard technique in most TV serials), as the viewers wait to see if and when the character will discover what they happen to know. A classic example is in Sophocles' *Oedipus Rex*, where the audience knows about Oedipus' parentage while Oedipus continues blindly on, unaware of the truth—and blinding himself when he discovers it. Turning to Shakespeare, the audience knows that Juliet is asleep, whereas Romeo, believing she's dead, kills himself, doubling the audience's chagrin. There can be foreshadowing involved as well, as when King Duncan declares that Macbeth has his trust, while the audience knows Macbeth is planning to kill him. Zero-focalization is the easiest way to provide this privileged information to the audience, but dramatic irony can be created with the other two methods as well.

However, focalization is hardly as simple as Genette's definitions suggest. As Edmiston (1991) points out, a first-person narrator who writes retrospectively about his younger 'experiencing' self could be viewed as internally-focalized because he is limiting himself to the point-of-view of that younger person; yet he not only knows more, but usually says more, expressing zero-focalization; finally, he can also act merely as a witness when describing other characters, thus expressing external focalization. In short, these three cases need not be mutually exclusive. A good example of all three focalizations is seen in Example 3.5, an excerpt from Frank O'Connor's autobiographical story, *The Genius*:

Example 3.5
Now one of the things I wanted badly to know was where babies came from, but this was something that no one seemed to be able to explain to me.

Skilled writers will often vary the focalization, as can be seen in James Joyce's *The Dead*, where as the story begins we hear the narrative voice of the maid, and then floating through different focalizations as the story weaves through the speeches and conversation of other guests at the party. Attempting to annotate the focalizations of sentences in this story suggests that focalization should really be viewed as involving multiple dimensions that can

3.2 Narrator Characteristics

coexist and blend together, as in Example 3.5 above—with zero-focalization providing a base or launching pad for excursions, of varying lengths, into other focalizations.

How does focalization affect narrative comprehension? A study by Jin et al. (2024) on Chinese narratives suggests that focalization shifts (e.g., from external to internal focalization or vice versa, or shifts from one character's perspective to another) demand more cognitive effort to process. The cognitive effort was measured by eye-tracking as well as a sentence-picture verification task, where participants are asked to judge whether a picture matches the semantic context of the sentence they have just read.

3.2.4 Focalization with LLMs

Can today's Generative AI be used to explore or even detect focalization?

Let's consider a crowdsourced story, from the Mostafazadeh (2016), shown italicized here in Example 3.6. For starters, what does GPT-4o make of this story?

Example 3.6

> **User Prompt**: Here is a story. Does it make sense? If so, why? If not, why not?
> *Tina made spaghetti for her boyfriend. It took a lot of work, but she was very proud. Her boyfriend ate the whole plate and said it was good. Tina tried it herself, and realized it was disgusting. She was touched that he pretended it was good to spare her feelings.*
> **Machine**: Yes, the story makes sense, and here's why: ...
> Human behavior: It's common for people to try to spare the feelings of someone they care about, especially in situations where they've put in effort. The boyfriend's behavior, eating the food and saying it's good, fits within typical human behavior patterns of wanting to avoid causing emotional pain.
> Tina's realization: After tasting the food herself and finding it unpleasant, Tina's understanding of her boyfriend's actions adds a layer of depth to the story. It suggests she recognizes his kindness, even though he may have been less than honest. This creates a subtle, heartwarming moment that builds on the theme of affection and understanding in relationships ...

The machine judges the story to be plausible, and suggests, as a human reader would, that whether the boyfriend liked the food or not, he was being kind, as part of a human behavior pattern of avoiding hurting someone one cares about.

What is the focalization in this story? This might be judged by Genette as being internally-focalized, as it expresses things from Tina's standpoint. Why so? Well, we have the phrases and words "she was very proud", 'realized', and 'touched' that could indicate it's internally-focalized or else omniscient. But once we read that "she was touched that he pretended", we know that the pretense is Tina's inference.

To keep things simple, let's go back to a tripartite distinction among focalizations, and explore this distinction using LLMs.

3.2.4.1 Zero Focalization

For zero-focalization, let's edit the internally-focalized story by hand to change the narrator perspective in the original text to be omniscient. The change is shown here in boldface (though it is in plain text when fed to the machine). Let's then ask the machine whether the story is plausible or not, requiring it to explain its answer.

Example 3.7

> **User Prompt**: Here is a story. Does it make sense? If so, why? If not, why not?
> *Tina made spaghetti for her boyfriend. It took a lot of work, but she was very proud. Her boyfriend ate the whole plate and said it was good. Tina tried it herself, and realized it was disgusting.* **Not knowing her boyfriend's tastes,** *she was touched that he pretended it was good to spare her feelings..*
> **Machine**: The story makes sense and captures a relatable situation. It highlights a common social interaction where someone might pretend to enjoy something in order to spare another person's feelings. The logic is coherent: Tina works hard on something (the spaghetti), her boyfriend says he enjoys it, but when Tina tries it herself, she finds it disgusting. Based on her own judgment, she assumes her boyfriend must have lied to protect her feelings, interpreting his action as kind.
> However, this relies on the assumption that both parties have similar taste preferences, which may not always be true. While Tina finds the spaghetti disgusting, her boyfriend might genuinely like it, meaning that her assumption could be wrong. In that case, he might not be pretending at all, and her interpretation of his intentions would be inaccurate.

GPT-4o correctly judges the revised story to be plausible, based on the logic, but based on the injected phrase establishing omniscience, it is alerted to the possibility that different people's taste preferences might vary. It has not been instructed as to the tripartite distinction, nor as to the fact that this version expresses omniscience, but it is able to react to the latter.

3.2.4.2 External Focalization

So far, we have looked at LLMs (which of course generate text in response to prompts) carrying out analyses of focalization. How about having them generate focalized narratives? For external focalization, let's get the machine to rewrite the omniscient version of the story so it is externally focalized along the lines of Hemingway's style. To do so, we need to first explicitly teach it a tiny bit of narratology by instructing it as to the definition of external focalization. GPT-4o generates a plausible and improved version, as shown in Example 3.8.

Example 3.8

> **User Prompt**: External focalization is when the narrator says less than the character knows. It is exemplified in Hemingway's behavioristic narrative, such as his story "The Killers". Here is a story. Please rewrite it so that it's externally focalized, where the narrator doesn't reveal what the character knows.
> *Tina made spaghetti for her boyfriend. It took a lot of work, but she was very proud. Her boyfriend ate the whole plate and loving the taste of it, said it was good. Tina tried it herself, and realized it was disgusting. Not knowing her boyfriend's tastes, she was touched that he pretended it was good to spare her feelings.*
> **Machine**: *Tina placed the plate of spaghetti in front of him. He picked up his fork, twirled it around the pasta, and took a big bite.*
> *"Not bad," he said, finishing the plate quickly.*
> *Tina hesitated, then picked up her own fork and took a bite. She stopped chewing, her face still as she lowered the fork back to the plate.*
> *"Really, it's good," he said, leaning back in his chair, hands on his stomach.*
> *She didn't say anything, just stared at the empty plate.*

The above example shows that the LLM has enough of a grasp of focalization that it can use it to rewrite very short stories accordingly.

3.2.4.3 Internal Focalization

For internal focalization, consider attempting to rewrite the story along the lines of Woolf's stream-of-consciousness style in *The Waves*. GPT botches it, but Claude 3.5 Sonnet generates a stylistically awkward version (note the transitive verbs 'consumed' and 'tasted' with dropped object arguments). Its generated output is, however, still plausible as a story, as shown in Example 3.9.

Example 3.9

> **User Prompt**: Here is a story:
> *Tina made spaghetti for her boyfriend. It took a lot of work, but she was very proud. Her boyfriend ate the whole plate and loving the taste of it, said it was good. Tina tried it herself, and realized it was disgusting. Not knowing her boyfriend's tastes, she was touched that he pretended it was good to spare her feelings.*
> Rewrite it to the same length to imitate the narrative style of Virginia Woolf's *The Waves*.
> **Machine**: *Tina's hands trembled, pasta-flecked. "I am creation", she thought. "I am pride". Her boyfriend consumed, each bite a silent homage. "I am acceptance", his mind whispered. "I am love". Tina tasted, reality stark on her tongue. "I am disappointment", she realized. "I am touched". The evening pulsed, a rhythm of unspoken truths, of separate consciousnesses merging in the act of giving and receiving, bound by strands of spaghetti and the tender fabrications of the heart.*

It would be incorrect to infer from this that Claude understands internal focalization as Genette does; rather, it has mastered Woolf's style and can mimic it. That, however, is sufficient qualification for humans to be able to use it to alter the focalization in a story.

3.2.4.4 Mixed Focalization Violations

Rather than considering the positive examples of mixed focalization, I focus here on cases where mixing is impossible. None of the tools I tried were able to detect that fusing the Woolf and Hemingway styles in the same sentence would be impossible. The best they could do was a mishmash, producing some sentences in the Woolf style and some in the Hemingway style.

Along the lines of detecting violations, let's introduce by hand an incoherent point-of-view (or pov) shift, such as from third person to first, as shown in Example 3.10.

Example 3.10

> **User Prompt**: Here is a story. Does it make sense? If so, why? If not, why not?
> *Tina made spaghetti for her boyfriend. It took a lot of work, but she was very proud. Her boyfriend ate the whole plate and said it was good. I tried it, and realized it was disgusting. I was touched that he pretended it was good to spare my feelings.*
> **Machine**: The story is coherent and makes sense, but there is a subtle issue with the perspective in the final sentence that can cause confusion. ...
> Perspective Issue: The final sentence introduces an ambiguity in point of view: The narrator says, "I was touched that he pretended it was good to spare my feelings." Since

3.2 Narrator Characteristics

> Tina is the one who made the spaghetti, it would make more sense if the boyfriend was sparing Tina's feelings, not the narrator's. If the narrator didn't make the spaghetti, there's no reason for the boyfriend to be concerned about their feelings.
> <u>Revision Suggestion</u>: If you meant that the boyfriend was sparing Tina's feelings, a minor revision would clarify that: "I was touched that he pretended it was good to spare her feelings." Alternatively, if you want to keep the narrator involved more directly in the situation, you could add some detail about the relationship between the narrator and Tina.

The machine notices the oddity of the pov shift, as well as the fact that the boyfriend should be sparing Tina's feelings rather than the narrator. It then suggests possible fixes.

3.2.4.5 Focalized Perceptions

One issue related to pov that has not been addressed that much in narratology is what a character can actually perceive, based on their physical situation. If the character is incapable of perceiving or acting in certain ways, the focalized narrative from their pov should sound implausible, or else suggest a plausible but unreliable narrator. Likewise, if part of a scene isn't perceivable by the character, any claims she makes about what's happening there should introduce some questions in the reader's mind. An example of the former (ability to recognize affordances of objects) is in Example 3.11, also from the (Mostafazadeh 2016). My addition is shown here (but not to GPT) in bold:

Example 3.11

> **User Prompt**: Here is a story. Does it make sense? If so, why? If not, why not?
> *There was a power outage in the Smith residence. The* **infant** *Fred could barely see in front of him. He crawled around in the dark, feeling his way around. Finally, he felt his hand wrap around a familiar item. He turned on the flashlight.*
> **Machine**: ... The story raises some issues with plausibility. While the sequence of events is logical, the fact that an infant, especially one young enough to be crawling, can recognize a flashlight and know how to turn it on feels unrealistic. Infants typically lack the fine motor skills and cognitive development needed to operate a flashlight. The story could work better if Fred were older, or if the focus was more on the sensory exploration than on him turning on the light.

The machine detects the implausibility of an infant's pov of a particular situation.

However, in other instances, the machine fails to detect the implausibility of perception (hearing at a distance) in Example 3.12, which is a hospital story adapted slightly from the ROC corpus.

Example 3.12

> **User Prompt**: Here is a story. Does it make sense? If so, why? If not, why not?
> *I was sent to the emergency room because I was cutting. They carried me on a bed from the ambulance, strapped in, and crying out in pain. The room they put me in was at the other end of the corridor from the nurse's station. Their door was closed, but they were laughing at me.*
> **Machine**: Your story conveys a powerful emotional scene, but there are some aspects that could use clarification to enhance coherence and impact. ...
> Potential Issues:
> <u>Laughter of the Nurses</u>: The line about the nurses laughing could be confusing or ambiguous. In real-life situations, it is unlikely that medical staff would laugh at someone in obvious distress. It could work as an unreliable narration, implying the narrator's perception rather than reality (if the narrator is in such mental pain that they interpret everything negatively). However, it would help to clarify if this is a subjective interpretation or an actual event.
> <u>Emotional Depth</u>: While the scene is striking, expanding on the character's inner experience: why they perceive the nurses as laughing, what kind of pain they are in (emotional, physical, or both) could make the narrative more layered and powerful.
> <u>Suggestions</u>: If the laughter is meant to be real, you might want to explain why the nurses are acting this way, as it seems implausible. Perhaps the narrator overhears something unrelated that they interpret as directed at them due to their emotional state. If it's meant to convey the narrator's sense of paranoia or heightened sensitivity, you could hint at their emotional state more clearly, maybe through internal monologue or physical sensations.
> In short, the story makes sense but could benefit from clarifying the relationship between reality and the narrator's perception.

Instead, the system focuses on another important issue with the story, namely the implausibility of nurses acting in that way and whether the event is merely the narrator's perception, as part of unreliable narration, or is real.

In Chap. 7, we will indicate how spatial reasoning from AI may be used in the future to infer what characters—or a camera narrator—can see based on their physical vantage point.

3.2.4.6 Focalization Summary

What have we found so far in our experiments on focalization? In terms of its adequacy in describing phenomena, the traditional three-way distinction appears overly simplistic. Focalization appears to have multiple dimensions that can coexist and blend together, with zero-focalization serving as a basic form. Using five-sentence stories from the ROC corpus, we varied the focalizations and assessed the plausibility, noting that LLMs are able to implicitly detect certain types of focalization and to detect it explicitly when instructed to. The machines are also, to their credit, able to rewrite stories to alter the focalization. However, they often struggle with mixed focalization.

3.3 Narrative Levels

3.3.1 Embedded Narratives and LLMs

Let us now return to the issue of embedded narratives. A story Y nested inside story X can be viewed as being at a higher *level* than story X. Genette (1980) defines this more precisely: "Any event a narrative recounts is at a diegetic level immediately higher than the level at which the narrating act producing this narrative is placed." The outermost level he calls *extradiegetic*, the first embedded level *diegetic* or *intradiegetic*, the second level as *metadiegetic*, the third level as *metametadiegetic*, and so on. This nomenclature is awkward, and we are better off counting levels, starting with the outermost level as level zero, so that if story X is at level n, a story Y nested inside it will be at level $n + 1$.

Perhaps the most celebrated example of embedded narratives is found in *The Thousand and One Nights* (TTaON). Many of the stories in TTaON contain multiple other stories, an example also known in the literature as a *frame story*. For example, Scheherazade's story *The Tale of the King, his Son, his Concubine, and the Seven Viziers* has an embedded story *The Lady and Five Suitors*, and her story *The Three Ladies of Baghdad* features the eldest of the three ladies (Amina by name) telling a story about being rescued by a magical woman.[3]

Another form of embedded narrative is for two narratives X and Y to embed a common third narrative Z. For example, narrative X could embed a popular Z tale and then narrative Y could have a character repeat that tale, either through narrating the entire Z (copying it verbatim), or else by reference, using a sentence like "She then repeated to the Sultan the story of the Fox and the Grapes." I will treat any repeating of $Z1$ by narrative Y as a distinct story $Z2$, irrespective of whether it is copied or referenced.

As indicated in Sect. 3.2.4, LLMs don't seem to have a problem with establishing narrator identity even for embedded narratives, thanks to their prowess in certain aspects of coreference. This extends to disambiguating names that 'shadow' the names in the embedding narrative. However, segmenting the entire structure of embedded narratives is a more challenging issue.

While one cannot test LLMs on the TTaON due to its length, its possible to construct a summary of a part of it that tests embedding of narratives. The input and output to GPT-4o is shown in Example 3.13.

---

[3] The compendium of stories that constitutes TTaON has no single author, and has evolved considerably over the centuries.

Example 3.13

> **User Prompt**: Z, a character in a story called TryMe, tells her boss Y a story called Three Ladies which features the eldest of the three ladies (X by name) telling a story about a magical woman P.
> Show the structure of the story TryMe. Keep your output short.
> **Machine**: The structure of the story TryMe can be represented as follows:
> Story A (TryMe): Z tells Y a story.
> Story B (Three Ladies): Z's story, featuring three ladies.
> Story C: The eldest lady (X) tells a story about a magical woman (P).
> This forms a nested structure with three narrative layers: TryMe contains Three Ladies, which contains the story of P told by X.

Given that the machine has no problem with this, where the structure is in effect revealed to the machine, let's make it do more work to infer the structure. Here is my constructed example that may have amused Claude Shannon, who worked on telephony. Both Gemini Advanced and GPT 4-o came up with comparable answers, and Gemini's is shown here:

Example 3.14

> **User Prompt**: What is the structure of this story?
> *Ali said to Jane "I asked Jill to call Kathy and give her instructions on how to explain to David why you broke up with him. Instead Kathy told David to ask Jill to do it, and Jill instead told David that she suspected you and I were having an affair while you were still dating David."*
> **Machine**: This story uses a nonlinear structure . . .:
> <u>Frame Story</u>: The outermost layer is the speaker telling us about a situation involving themselves, you, and David. This sets the stage but doesn't give us the whole story directly.
> <u>First Embedded Narrative</u>: Within that frame, we hear about the speaker asking Jill to do something (call Kathy). This is our first layer down.
> <u>Second Embedded Narrative</u>: Jill was supposed to tell Kathy to do something (explain to David). This is another layer deeper, like a set of Russian nesting dolls.
> <u>Third Embedded Narrative (and Twist)</u>: Instead of following the plan, things go awry. Both Kathy and Jill deviate, with Jill ultimately creating a false narrative about an affair. This is the innermost story and the one with the most impact.
> This structure creates dramatic irony and suspense. We know the speaker's intention, but we see how the message gets distorted with each retelling. It also highlights the unreliability of communication and the potential for misunderstandings.

> Think of it like a game of telephone, where the original message gets completely changed by the time it reaches the end. This structure keeps the listener engaged because they are trying to piece together what really happened and what the consequences of this misinformation will be.

More complex narrative embeddings over such short stretches do not fare well with either humans or machines, due no doubt to different kinds of memory limits.

3.3.2 Narrative Threads

Conversations can of course have multiple threads or topics, which can sometimes be hard to disentangle. There is thus quite a lot of research still on disentangling threads in chat. It also not uncommon to come across tangled threads in narrative, but here the conventions are such that the signals of a new thread are often clear to a human. We already saw the adult and child perspectives being mixed mid-sentence in Woolf's *Mrs. Dalloway*. Here is an example from William Faulkner's *The Sound and the Fury* where the boundary between narratives, in this case too by a first-person narrator who is an adult and the same narrator as a child, occurs abruptly without explicit marking.

Example 3.15
We went along the fence and came to the garden fence, where our shadows were. My shadow was higher than Luster's on the fence. We came to the broken place and went through it. "Wait a minute." Luster said. "You snagged on that nail again. Can't you never crawl through here without snagging on that nail." Caddy uncaught me and we crawled through. Uncle Maury said to not let anybody see us, so we better stoop over, Caddy said. Stoop over, Benjy. Like this, see. We stooped over and crossed the garden, where the flowers rasped and rattled against us. The ground was hard.

The reader may have to process further down in the narrative (excluded for reasons of space), or read it several times, to determine that the boundary between the adult and child narrative is after the word 'nail'. The LLMs don't nail the boundary at all, but Faulkner and his publisher were kind enough to italicize the flashback which begins after that word. The child narrative is also marked by a transition from Direct Speech (i.e., with quotes, e.g., "*See them ...*") to a mix of Reported Speech (*Stoop over, Benjy.*) and the Free Indirect Style (*We stooped over ...The ground was hard.*)

The disentangling of narrative threads in stories is an interesting challenge for machines. Work by Wallace (2012) has shown moderate success in disentangling multiple narrative threads in the novel *Infinite Jest* by David Foster Wallace, achieving an F-measure of 0.7 using supervised machine learning.

The F-measure, by the way is a metric for accuracy that can be practically more useful than simple accuracy (the latter measured in terms of percentage correct). Let's say an AI system (looking, say, at patient data) bins examples into two classes: positive and negative. Typically, we want to know when the system guesses positive, whether it's correct. (After all, a false positive—flagging an example as positive when it's not—can be really misleading and have all sorts of unintended consequences.) The Precision measures exactly that. Also, we want to know how many of the positive examples did the system actually find in the examples it marked as positive. The Recall measure does that. It would be ideal to have a single measure that takes both Precision and Recall into account, and the F-measure does that using the (harmonic) mean of Precision and Recall, called F_1 in Eq. 3.1 below. The details can be skipped if desired.

$$\begin{aligned} \text{Precision} &= \frac{TP}{TP + FP} \\ \text{Recall} &= \frac{TP}{TP + FN} \\ F_1 &= 2 \cdot \frac{\text{Precision} \cdot \text{Recall}}{\text{Precision} + \text{Recall}} \\ &= 2 \cdot \frac{TP}{2TP + FP + FN} \end{aligned} \quad (3.1)$$

where:

TP = True Positives (correctly predicted positives)

FP = False Positives (incorrectly predicted as positive)

FN = False Negatives (missed positive cases)

3.3.3 Metalepsis and LLMs

Metalepsis is a transgression between what is told and its telling. As defined by Genette (1980), it is a narrative strategy where (a) the narrator intrudes, with or without the reader, into the narrative (i.e., into the diegetic universe) or (b) where diegetic characters extrude into the metadiegetic universe.[4] Metalepsis of type (a), with narratorial intrusions is rather more common than type (b).

Metalepsis has attracted considerable narratological attention. The narratologist (Pier 2014) argues that it applies only to fiction, and notes that in addition to the transgressive merging of multiple levels of narrative, "putting the story time on hold story time while the narrator, in a relative cohabitation with the character, intervenes with a metanarrative comment." (Pier 2014, p. 4). Pier also suggests that metalepsis is part and parcel of a larger strategy of *defamiliarization*, a concept introduced by Shklovsky (1973). The latter proposes

---

[4] Genette again diverges from the terminology of classical rhetoric, where *metalepsis* means a metonymy involving a double substitution, e.g., "ears of corn" (Quinn 1982).

3.3 Narrative levels

that art must impart the sensation of things as they are perceived, rather than how they are known. As such, the contents of the fabula must be rendered in the discourse in a way that makes them unfamiliar, prolonging the processing of the text by the reader for aesthetic purposes.

As a prototypical example of defamiliarization, Shklovsky cites Lawrence Sterne's *Tristram Shandy*, a novel we mentioned earlier that is famous for its narrator's intrusions, involving metalepsis of type (a), digressions, and temporal disruptions. Pier argues that defamilarization works to convey to the reader "that art is 'made' of devices" (Pier 2014, p. 15).[5]

Metalepsis of type (b) is exemplified in Julio Cortazar's short story *The Continuity of Parks*, which allows a character and his actions to extrude into the extradiegetic world. Here is an excerpt from Cortazar (1967), with my elisions indicated by ...:

Example 3.16
He had begun to read the novel a few days before. ...That afternoon, ...he returned to the book in the tranquility of his study which looked out upon the park with its oaks. Sprawled in his favorite armchair, its back toward the door—even the possibility of an intrusion would have irritated him, had he thought of it—he let his left hand caress repeatedly the green velvet upholstery and set to reading the final chapters. ...Word by word, caught up in the sordid dilemma of the hero and heroine, letting himself be absorbed to the point where the images settled down and took on color and movement, he was witness to the final encounter in the mountain cabin. The woman arrived first, apprehensive; now the lover came in, his face cut by the backlash of a branch. ...The estate manager would not be there at this hour, and he was not there. He went up the three porch steps and entered. ...At the top, two doors. No one in the first room, no one in the second. The door of the salon, and then, the knife in hand, the light from the great windows, the high back of an armchair covered in green velvet, the head of the man in the chair reading a novel.

In this story, the third-person narrator (call him $n1$) describes a character (call him A) reading a novel X involving a character B who eventually sneaks up on A to kill him.

The implications of metalepsis for temporal analysis are discussed in Chap. 5.

Here is metalepsis (of type (b)) applied by Gemini Advanced to one of the ROC stories we encountered earlier:

---

[5] Confining ourselves to fiction again, Piers' view seems consonant (to me) with the claim by McHale (2003) that while 'modernist' fiction interprets the world and our relation to it, 'post-modernist' fiction lays bare the scaffolding underlying the literary creation.

Example 3.17

> **User Prompt**: Here is a story. Add a short ending sentence to it that displays metalepsis and a twist.
> *There was a power outage in the Smith residence. Fred could barely see in front of him. He crawled around in the dark, feeling his way around. Finally, he felt his hand wrap around a familiar item. He turned on the flashlight.*
> **Machine**: *... and illuminated the author, who was frantically scribbling the final words of the story by the light of his laptop screen.*

3.3.4 Subordinated Discourse

Speech acts of narration, including showing and telling and saying, whether embedded or not, as well as mental events such as dreams, memories, imaginings, reveries, states of believing, etc., are more generally treated as varieties of **subordinated discourse**. While embedding as discussed above involves subordination through narration, mental events introduce multiple planes of discourse. In Proust's *Jean Santeuil*, analyzed by Genette (1980), we have Jean's memories (of a past involving rainy days with his nursemaid), and his thoughts (at that past time) of the future, as shown in Example 3.18:

Example 3.18
Sometimes passing in front of the hotel he remembered the rainy days when he used to bring his nursemaid that far, on a pilgrimage. But he remembered them without the melancholy that he then thought he would surely some day savor on feeling that he no longer loved her.

Subordinated discourses do not always involve embedded discourses per se. However, the contents of the memories and thoughts are distinct from actual happenings. They fall under what has been called *speculative language*, as discussed by Sauri et al. (2008). When characters express beliefs, memories, etc. via subordinating contexts, the information introduced in those contexts is treated as non-factual. So, if John remembers Mary hitting Bill, that event of Mary hitting Bill will be treated as non-factual. But if John knows that Mary hit Bill, then Mary hitting Bill is factual. Thus, we must test the predicate involved in the subordinating context.

Accordingly, the narrative model will need to distinguish between factual and non-factual events; further, the scope of the subordinating event will have to be indicated. In "John believed the man was armed. He decided to hand over the money," the second sentence is outside the scope of John's believing. And in Example 3.19 from Sauri et al. (2012), the event of pretending and the state of being safe are subordinated to the saying, and are therefore hypothetical. The proposition that nuclear power is safe is presented as counter-factual by

3.3 Narrative levels

source Nelles, and factual by Germany, according to the narrator, whose own position is left underspecified.

Example 3.19
Nelles said that Germany has been pretending for long that nuclear power is safe.

Of course, an embedded narrative may have a subordinated discourse, just like any other narrative; further, subordination can occur in factual as well as fictional narratives. From a semantic point of view, the subordinated references can be analyzed, if desired, in terms of possible-world semantics using modal logic as defined by Kripke (1959). We will have more to say about this in Sect. 3.3.5.

> **Subordination by unreliable narrators**
> We already came across multiple instances of this pattern, including *The Murder of Roger Ackroyd*. To turn to an even more vivid example, in Ron Howard's film *A Beautiful Mind*, several of the key scenes and characters the audience witnesses are later revealed to be hallucinations on the part of the principal protagonist (the character John Nash), changing their fictional status. Thus, from a narratological standpoint, the audience must initially be led to believe that the character Charles Herman (a literature student) has an independent existence, albeit in a fictional world, outside the beliefs of Nash, whereas eventually the audience must revise that belief about his ontological status; the same is true of the character Parcher. Unreliable narration can thus be a trigger for *belief revision*. It is worthwhile noting that the revision in this case involves the reader realizing that the narrator was presenting a focalized view (from Nash's perspective) without cuing the viewer about it.

3.3.5 Accessibility

Let us turn to a question we have delayed addressing until now. What do characters in narratives know and believe? How does this relate to embedded narratives and subordinated discourse? To address this, we need to introduce the notion of *Accessibility*.

> **Accessibility in modal logic**
> In modal logic, which we mentioned when discussing approaches to meaning in Chap. 1, the truth of propositions is qualified by notions like *possibility* and *necessity*; these notions are used to provide a semantics for natural language sentences that involve references to the knowledge of agents, such as "John knows that Mary hit

> Bill." The classic treatment of modality from Kripke (1959) is that an agent knows something if it is true in all situations that she deems possible; these situations are called **possible worlds**. The worlds that are possible for an agent are determined by an **accessibility** relation that defines which situations are possible for the agent given each situation the agent can be in. Thus "John knows that Mary loved Bill" is necessarily true in the actual world if and only if "Mary loved Bill" is true in all possible worlds corresponding to John's knowledge, i.e., in all possible situations that he knows about.

We now introduce some logical background on accessibility which may be skipped if desired.

As noted by Halpern et al. (1992), the limits to an agent's knowledge fall out of the properties of the accessibility relation. A reflexive accessibility relation R (so that for every x, $R(x, x)$) means that the actual (real) world is one of the possible worlds, i.e., the agent cannot know things that are false. A transitive accessibility relation (i.e., for every x, y, z, $R(x, y)$ and $R(y, z)$ implies $R(x, z)$) means that the agent knows whatever she knows: she is omniscient to the extent of knowing all logical implications of everything she knows. If R is both transitive (as above) and symmetric (i.e., for every x, y, $R(x, y)$ implies $R(y, x)$), then she knows what she doesn't know. The accessibility relation thus has to be constrained so as to limit the knowledge of agents to what is realistic.

Modal logic is naturally extended from representing the meaning of sentences involving references to knowledge to sentences that involve references to belief, imagination, memory, obligation, future and past, etc. Thus, when considering Example 3.18, the proposition "Jean remembered that he and his nurse went on a pilgrimage" is necessarily true in the actual world if and only if "he and his nurse went on a pilgrimage" is true in all possible worlds corresponding to his memories.

In the **ontological promiscuity** approach of Hobbs (1985), one can proliferate entities (in a 'Platonic' universe), but restrict **existence** to those entities that are actual. To say that John runs, in Hobbs's account, means that the entity of John's running exists in the actual world. When we say that John wants to fly, it will mean that John and his wanting exist in the actual world, but his flying does not exist in the actual world. Likewise, to say that Ravi worships Vishnu will mean that Ravi and his worshiping exist in the actual world, but Vishnu does not.[6]

In my approach, **accessibility** is constrained by two conventions:

1. the **embedding norm**: a character in an embedded narrative cannot access one from an embedding narrative, unless the latter is an importable character, such as generally familiar places or characters;
2. the **knowledge norm**: if someone doesn't know of an entity, it is inaccessible to her.

---

[6] This idea is also developed substantially in the philosophy of language, most recently by Parsons (1980), who defines the concept of a set of individuating or *nuclear* properties that characterize fictional entities.

Thus, in the TTaON, while Shahryar has Scheherazade in his accessibility set, Amina does not. Likewise, any entities believed by Amina are only in Amina's accessibility set. All of these entities, being fictional, have a flag, based on ontological promiscuity, called `exists`. If it is a fictional entity, it is set to false. Note that if a real character is declared by a non-fiction narrator to be lying about or imagining some entity, that entity will be subordinated. If it is not declared, then it will have its exists flag set to true. This flag setting is defeasible, as in the case of unreliable narrators, or where we later discover that the character was lying earlier, or mistaken due to perceptual or other limitations.

To exemplify further, in *A Beautiful Mind*, the story world is peopled with entities such as Nash, Charles, Alicia (whom Nash marries), Parcher, and the place Princeton. All of the people have their `exists` flag set to false; as for Princeton, assuming it is the real Princeton and not a surrogate (see Parsons 1980), its `exists` flag is set to true. All four of these entities are accessible to Nash. When Charles is revealed to be a figment of Nash's imagination, the audience realizes that Charles was never accessible to Alicia. Thus, the audience model may have initially assumed Charles being accessible to Alicia; the revision involves deleting Alicia from Charles's accessibility set.

As another example, in Cortazar's story *The Continuity of Parks*, where a character A reads a novel involving a character B who eventually sneaks up on A to kill him, B is accessible to A, but given the embedding norm, A is not accessible to B. So the sneaking up involves making A accessible to B, which involves a revision of belief that violates the embedding norm.

Computing such accessibility relations does not, thankfully, require modal reasoning. It requires determining the textual scope of intensional predicates, as well as taking into account the type of focalization.[7] In language understanding, flagging such predicates is easier when they are explicitly mentioned. In cases involving belief revision, the computation is more complex (and may involve, if logic-based, non-monotonic reasoning).

3.4 Character Representations

While references to characters' goals and motives are crucial for explaining the motives behind actions, modern humanities narratology has typically refrained from creating formal models of characters (or narrators) and their goals. One need not look far for justifications for this attitude. As Genette (1988) argues, the study of characterization is not the proper province of narratology:

> In my view it is decidedly, although relatively, preferable (more "narratological") to decompose the study of "characterization" into the study of its constituting devices (which are not specific to it): denomination, description, focalization, narrative of words or thoughts or both, relation to the narrating situation, etc.

---

[7] The problem can be viewed as similar to that of event factuality tagging of Sauri et al. (2012).

A little further on, Genette goes on to say that reader response is also outside the scope of narratology (p. 153):

> Sympathy or antipathy for a character depends essentially on the psychological or moral (or physical!) characteristics the author gives him, the behavior and speeches he attributes to him, and very little on the techniques of the narrative in which he appears.

Both these quotations are offered by Genette in defense of his focus on form, not content.

The AI perspective on characters in narrative is quite different. Formal representations of characters as intentional agents are critical for any narrative understanding system that attempts to answer questions about characters' motivations, as well as for narrative generation systems that have to create interesting behaviors for characters. Likewise, representing the beliefs of the audience and readers about characters, which is part and parcel of modeling audience response, is important in narrative understanding systems, and crucial for improved narrative generation and interactive narrative. The fact that such characterization pertains to content rather than form is not an impediment to AI modeling. Nevertheless, the reasoning that goes on in the plans of characters in AI (discussed later at some length in Chap. 6) relate to the fabula, and as such are of less interest to narratology than the impact any such reasoning may have on the structure of narrative discourse. Thus, for character plans and goals to have relevance to humanities narratology, the impact of any character's plans on the discourse is an important issue.

The narratological properties of characters have several different facets:

Traits

We already came across traits in Chap. 2 when discussing marriage proposals involving Ali and Jane. In general, characters are distinguished by attributes related to their personality, appearance, ethical values, etc., and these are drawn from a relatively small inventory. Some character traits form clusters, or stereotypes, as in stock characters like the *femme fatale*.

Roles

Since our inventory of emotions and personalities is limited, no matter how creative a writer may be, the characters she introduces have to conform to a set of functional or formal roles. In folk tales, the roles of the hero, villain, and helper may be universal across cultures (these are examples from Vladmir Propp's classification, discussed in detail later in Chap. 4). In other genres, the roles are of course varied, but still relatively small in number, perhaps limited to a few hundreds.

Evolution

Characters can also inherit characteristics bequeathed by event outcomes, as in the case of a spy or princess who becomes a prisoner, with her new role integrated in some way with the old. As characters evolve, their properties can change, and those changes and transformations are what help make stories of relevance to human societies.

Narrative Attention
There is also a distinction to be made as to what sort of attention the narrator pays to the characters. In discussing the attributes, Forster distinguished between 'flat' and 'round' characters (Forster 1956). Flat characters, as in many of Dickens' characters, are "constructed around a single idea or quality", meaning they are shallow, or one-dimensional stereotypes, whereas round characters are "capable of surprising in a convincing way", meaning that they have more depth or dimensions, in other words, a more varied set of attributes. A judicious combination of the two types, Forster points out, is in fact to be found in Dickens' work. Often, a flat character will become far more rounded by the time the story ends, as in Flannery O'Connor's story *A Good Man is Hard to Find*, where the Grandmother character is transformed as she is facing death at the hands of the Misfit, a hardened killer who also evolves at that moment.

Geographic and Diachronic Evolution
The notion of character need not, of course, be text-specific. Characters can be tracked both geographically and temporally, revealing how different cultures and ages present and react to them. For example, Robin Hood starts out as an anti-clerical outlaw who robs the rich to help the poor, then, in the 19th Century, becomes a regional hero battling Norman nobility, and winds up as a fox in a Disney film, and no doubt another future awaits him as an AI character. The character Krishna appears in many ancient tales from India, and readers often have prior assumptions about well-known characters like him; these are refined and elaborated as the characters are encountered in new incarnations.

Turning to computation, it is less interesting to classify characters into different types than to cluster them and see what emerges. By carrying out dependency parsing and coreference across entities, research by Bamman (2014) has been able to mine a database of more than 15,000 novels to test 30 literary hypotheses. They use a Bayesian statistical model that can predict different character types, using features such as the actions that a person participates in, the objects they possess, and their attributes. The system is able to identify cases where two characters by the same author happen to be more similar to each other than to a closely related character by a different author. For example, they confirm the hypothesis that Wickham in Jane Austen's *Pride and Prejudice* (1813) resembles Willoughby in her *Sense and Sensibility* (1811), more than either character resembles Mr Rochester in Charlotte Bronte's *Jane Eyre* (1847). The machine also confirms that Elizabeth Bennet in *Pride and Prejudice*, resembles Elinor Dashwood in *Sense and Sensibility* more than either character resembles Elizabeth's marriage-obsessed mother, Mrs Bennet.

Systems can also analyze character relationships, tracking their evolution, as Iyyer et al. (2016) have done, mining nearly 1400 works of fiction in an unsupervised manner. For each pair of characters, their program identifies a sequence of fixed-length spans of text that mention them. The spans are represented in terms of three kinds of embeddings: (i) word embeddings (ii) character embeddings, which their neural net with fill with immutable aspects of characters across the work (e.g., Javert in *Les Miserables* is a police officer), and

(iii) book embeddings that will be eventually filled with global context about books (e.g., *Moby Dick* takes place at sea). By analyzing how these descriptors shift over the course of the narrative, the system constructs a *relationship trajectory*. This approach is able to infer, for example, the correct trajectory of the relationship between Arthur and Lucy in the novel *Dracula*: it starts with 'love' (which is a label given by the human evaluator to a set of highly weighted words) and ends with 'murder' (likewise). However, the output is numeric (namely, word vectors) which have to be interpreted by humans, though potentially it could be fed to an LLM for sentence generation.

Of particular interest to humanities narratology is the character-related work of Elson (2010). They automatically tag quoted speech (an aspect of narrative distance), attributing quotes to characters, and building a graph marking length of conversations between pairs of character entities. Their automatic detection and analysis of such conversational social networks in 19th century novels provides evidence against a claim by the literary critic (Moretti 1999) that there is an inverse correlation between the amount of dialogue and number of characters in such novels.

3.5 Audience

3.5.1 Preliminaries

All narratives are consumed by some **audience**, whether **reader**, spectator, hearer, or sets of these. In some narratives, the audience is referred to explicitly, as in invocations to "the reader". There is also the second-person "you" that refers to the audience, as in the opening sentence of J. D. Salinger's *The Catcher in the Rye*, shown in Example 3.20.

Example 3.20
If you really want to hear about it, the first thing you'll probably want to know is where I was born, and what my lousy childhood was like and how my parents were occupied and all before they had me, and all that David Copperfield kind of crap, but I don't feel like going into it.

In other narratives, the audience may be created by the narrative, as in the case of King Shahryar being the audience for Scheherazade's narratives.

The audience may sometimes be treated as if it is an active participant in the narrative, as in Sterne's *Tristram Shandy*: "I beg the reader will assist me here, to wheel off my uncle Toby's ordnance behind the scenes." Here the reader is being invoked through a subordinated discourse mediated by "I beg." Sometimes, the audience of an embedded story may interact with the narrator, as is often the case in Scheherazade's embedded stories. In this case, the audience is just another character or set of characters. This interaction is also found in traditional oral storytelling. In interactive narrative systems, the user/player is usually a co-author of the story, and can also be its audience, along with other, passive readers.

A reader usually develops beliefs about different characters in the story, or even of the narrator. In the latter case, we have what narratologists, following (Booth 1983), call the **implied author**. That is the impression of the author created by the work, as opposed to the impression of the author created from any background information about the author. The distinction between the implied author and the narrator is far from clear, and Genette (1980) does not represent this distinction, citing in his defense Occam's Razor ("Entities are not to be multiplied without necessity"). For computational purposes, the implied author is hard to model and I will ignore it. This is not to say that fine-grained distinctions are not relevant; a more detailed user model may make such a distinction, though the yardstick for it will have to be made clearer.

The **audience** in the sense in which I use the term is also called the **actual reader** in narratological studies. I do not model various finer-grained narratological notions, such as the **implied reader** of Iser (1974), the addressee or **narratee** of Prince (1982), or the **authorial audience** of Rabinowitz (1988), who accepts "the author's invitation to read in a particular socially constituted way that is shared by the author and his or her accepted reader" (ibid., p. 22). These distinctions, while insightful in themselves, are open-ended and are hard to represent in a precise and formal manner for computational purposes, and are prone to be difficult to annotate. As a result, they are left out of our annotation scheme NarrativeML.

3.5.2 Audience Response

The audience naturally has beliefs about the narrator and characters in a work such as a novel. For example, if someone is a bank robber, the audience may assume, based on certain cultural stereotypes, that the character is violent. If someone speaks in gangster slang, the audience may believe he is a gangster. Background knowledge that may be culturally specific is often used in intelligent narrative systems that interpret narrative in a non-interactive way. From the standpoint of such systems, the audience *is* the interpreting system. In generating narrative, a **user model** (or reader or audience model) is highly desirable.

Readers (or members of the audience) may differ in terms of their construals of the beliefs and motives of characters in the narrative. The narrator may also ensure that the audience knows more about events than the character, leading to *dramatic irony*, which we discussed earlier under External Focalization.

The classic Aristotelian theory of audience response (Aristotle 1932) is based on the notion of *pity*. In his account, dramatic tragedy works on the audience by arousing emotions of pity (sympathy) and fear (antipathy). Pity is the feeling associated with a particular type of cognitive reasoning: the person having pity feels pain toward the pitied, who does not deserve the evil which has befallen him. The audience is able to empathize with a character by imagining the same experience happening to them.

While we are discussing the audience's response, it is worth listing the many ingredients in a narrative that can make the audience judge it to be interesting. These facets of **reader affect** include, as a laundry list:

- *Causal connection*: Are the events related to each other in plausible ways?
- *Pacing*: Does the narrative move along quickly, or does it dawdle? These perceptions of course depend on the stage of the narrative, and estimates of elapsed time in the story.
- *Suspense*: Are expectations set up which engage the reader? Is dramatic irony used effectively?
- *Resolution*: Is there a worthwhile ending?
- *Characterization*: Are the characters interesting, and believable?
- *Style*: Does it engage and stimulate?
- *Coherence*: Is it integrated into a whole?
- *Vividness*: Is it immersive and memorable? Does it invite you into a separate and distinct world?
- *Morality*: Does it grapple with moral issues, inspiring exemplary behavior or self-discovery?
- *Creativity*: Does it come up with interesting and surprising new developments?

As we shall see in subsequent chapters, causality, pacing, suspense, resolution, characterization, and coherence have aspects that can be computed. The other facets are, however, harder to model, given that many different parameters of a discourse are responsible for these audience impressions.

3.5.3 Evaluating Event Outcomes

It is undeniable that the reader's impressions about the motives and actions of an agent are formed against the background of a set of moral principles that the reader has adopted (or has been co-opted to adopt). Literary critics have homed in on the role of morality in readers' interpretations; as Rabinowitz (1988) argues,"there can be little doubt that the process of moral evaluation plays a central role in the reading of narrative fiction." A narrative, one might argue, makes a moral case for a given agent in terms of outcomes of events he or she is involved with. This has the effect in the reader's mind of possibly boosting or casting aspersions on that agent in terms of their moral character. These reader's reactions naturally also affect their emotions; as Hogan (2003) has suggested, narratives allow a reader to empathize with particular character, by priming their own emotions. In other words, we put ourselves in other's shoes, if only for a while.

To get a computational theory of audience reaction and satisfaction off the ground, it helps to simplify the reader's response in terms of a Boolean model of *emotion* involving judgments of sympathy or antipathy toward the character given the outcomes of events for the character. For example, in *Oliver Twist*, a reader might applaud when Fagin ends up hanged, and sympathize when Oliver is rescued by a rich man. Note that the reader is evaluating the character of Fagin in terms of the outcome of an event he is involved in, rather than the outcome for him. Thus, an outcome that is perceived as negative by Fagin may be perceived

3.5 Audience

as positive by the reader, and vice versa. We therefore need to measure agreement among readers on their character evaluations. High agreement will lay a foundation for reliable user models of actual reader responses.

While percentage agreement among raters is useful, the raters may guess, or always choose the class which has the most examples (the majority class). A measure called Cohen's *Kappa* corrects for such problems by comparing the agreement to chance agreement, as shown in Eq. 3.2. Its definition can be skipped if desired.

$$\kappa = \frac{p_o - p_e}{1 - p_e}$$

where:

$$p_o = \text{Observed agreement} = \frac{1}{n}\sum_{i=1}^{k} n_{ii}$$

$$p_e = \text{Expected agreement by chance} = \frac{1}{n^2}\sum_{i=1}^{k} n_{i\cdot} \cdot n_{\cdot i}$$

n = Total number of items
n_{ii} = Number of items both raters assigned to category i
$n_{i\cdot}$ = Number of items rater A assigned to category i
$n_{\cdot i}$ = Number of items rater B assigned to category i

(3.2)

To formally model character evaluations, I begin with a representation of the narrative fabula in terms of the events, their participant roles, and their underlying chronology. The sequence of events in narrated order is provided to the reader. Each reader rates each successive event outcome against each involved agent into exactly one of three categories: positive, negative, or neutral. These evaluations are marked on events on the discourse timeline. A positive evaluation (marked $++F$) means that the reader expresses sympathy with the agent F (irrespective of whether it is positive or negative for F), and a negative one (marked $-F$) means that she expresses antipathy toward F. Neutral evaluations are unmarked, unless they involve a change from positive or negative evaluations. Such a binary classification is admittedly crude, but it is a first step toward more polyvalent classifications of affect. Character evaluations are marked on the mentions of events, indicating that they are evaluations of the particular character based on the outcomes of such events.

Let us analyze character evaluations for Chekhov's story *The Lady with the Pet Dog*. Since the text is too long to include here, consider the following summary:

1. Gurov, a married man on vacation, seduced Anna, a married woman visiting Yalta.
2. Anna was ashamed that Gurov debased her. $-G$
3. When Gurov returned to Moscow, he was haunted by his memory of Anna.

4. He confronted Anna at a theater in her hometown S__, where he begged her to understand his love. ++*G*
5. She confessed how much she had longed for him. ++*A*
6. They met repeatedly in secret in a hotel room in Moscow, where they contemplated the great difficulties that lay ahead. ++*G*, ++*A*.

The annotations by an actual reader *R* are indicated alongside. *R* does not pass judgment on Gurov's act of seduction in part because the seduction is presented against a backdrop of scenes focalized through Gurov in a non-judgmental way so as to make the seduction entirely natural. However, once Anna's reaction is presented, *R* begins to dislike Gurov for his action. Another reader may of course have an entirely different opinion. Subsequently, when Gurov falls in love, his being haunted doesn't invoke sympathy from *R*, but the initiative he shows in seeking her out at the theater, and her positive response, results in *R*'s enhanced appreciation for each. The ending, where they carry out their furtive affair facing the troubles that lie ahead, is tragic, making *R* sympathize with their predicament.

Character evaluations are not typically annotated on summaries, however, but on individual sentences. Further, they are applied to elements derived from a semantic representation of the sentences. As an example, if the second element above in the summary was a single sentence, here is the semantic representation we assume for it, and the associated character evaluation.[8]

Example 3.21
Anna$_A$ was ashamed$_{e1}$ that Gurov$_G$ debased$_{e2}$ her.
EVAL(NEG, r, G, e2)

A similar experiment in character evaluation has been carried out by Elson (2012), where humans were able to reliably annotate, with high agreement, positive, negative, or neutral event outcomes for a given agent in the domain of Aesop's fables. Automatic annotation using lexical and syntactic features, however, did not perform above chance.

---

[8] The reader may be wondering how to read an expression like EVAL(NEG, r, G, $e2$). It means that the reader r annotated the character of Gurov as negative in relation to the outcome of the event of Anna being debased.

It is worth stressing here that my theory is obviously very crude:

- The audience may be sympathetic to a character like Oedipus, marking a positive character evaluation for him after he blinds himself; but the audience typically has a wider range of emotions than the Aristotelian dichotomy of sympathy/antipathy.
- We often hold conflicting emotions, just as we may be drawn to opposed goals from which we have to choose. To represent that, one could avail of a vector of emotions, with weights for different emotions, but that departs significantly from the goal of 'lightweight' annotation.
- Sometimes the narrator will make us self-conscious for having sympathized with a character whose current action or state makes us uncomfortable. Examples are ubiquitous, and include the final sentence of George Orwell's *1984*: "He loved Big Brother." We sympathize with Winston as a victim, but we do not approve of Winston's attitude—a point discussed at some length by Phelan (1989). So, our sympathy must not be confused with our approval. But marking the audience's beliefs and attitudes, while essential, has not been attempted in NarrativeML.
- Authors often strive to create discomfort during endings, using devices such as irony and ambiguity. Consider the beginning of the famous last paragraph of Thomas Hardy's *Tess of the d'Urbervilles*: "'Justice' was done, and the President of the Immortals (in Aeschylean phrase) had ended his sport with Tess." When reading that sentence and the rest of the paragraph, we reflect not only on the tragedy of Tess (and Lisa-Lu and Angel) but are disturbed by the narrator's apparent throwing up of his hands to fate, though his precise attitude is left ambiguous. Marking and evaluating such narrator attitudes is left to future work based on sentiment analysis (for example, using the approach of Pang et al. 2002).

3.6 NarrativeML

Based on what we have discussed so far, a basic annotation scheme can be defined in XML. We will assume that entities, events, times, and places are basic, and added to that are notions specific to narrative, pertaining to narrative, narrator, and audience. The annotation scheme has been used for marking up corpora, to support narrative understanding as well as generation. We will begin with a document-level annotation scheme called **NarrativeML**, expressed as an XML Document Type Definition (DTD), shown in Fig. 3.1. Although DTD's are rather primitive, they are quite transparent, and useful for specifying the annotation scheme. Next, we will show an example of linguistic data annotated with this scheme. All

```
<?xml version="1.0" encoding="UTF-8"?>
<!DOCTYPE NarrativeML [
<!ELEMENT NarrativeML (#PCDATA | NARRATIVE)*>
<!ATTLIST NarrativeML xsi:noNamespaceSchemaLocation CDATA #IMPLIED>
<!ATTLIST NarrativeML xmlns:xsi CDATA #IMPLIED>
<!ATTLIST NarrativeML version CDATA #FIXED "0.4">
<!ELEMENT NARRATIVE (#PCDATA | NARRATOR | AUDIENCE | CHARACTER | EVENT | TIME | PLACE | EVALUATION | SEGMENT)*>
<!ATTLIST NARRATIVE id ID #REQUIRED>
<!ATTLIST NARRATIVE parent IDREF #IMPLIED>
<!ATTLIST NARRATIVE narrator IDREF #REQUIRED>
<!ATTLIST NARRATIVE title CDATA #IMPLIED>
<!ATTLIST NARRATIVE medium CDATA #IMPLIED>
<!ATTLIST NARRATIVE level CDATA #IMPLIED>
<!ELEMENT SEGMENT (#PCDATA | TIME | EVENT | PLACE | SPATIALREL | TLINK | SLINK | MENTION)*>
<!ATTLIST SEGMENT id ID #REQUIRED>
<!ATTLIST SEGMENT title CDATA #IMPLIED>
<!ELEMENT NARRATOR (#PCDATA)>
<!ATTLIST NARRATOR id ID #REQUIRED>
<!ATTLIST NARRATOR exists (true | false) #IMPLIED>
<!ATTLIST NARRATOR type (present | absent) #IMPLIED>
<!ATTLIST NARRATOR accessibleTo IDREFS #IMPLIED>
<!ATTLIST NARRATOR name CDATA #IMPLIED>
<!ATTLIST NARRATOR form CDATA #IMPLIED>
<!ATTLIST NARRATOR order (ACHRONY | ANALEPSIS | CHRONICLE | PROLEPSIS | RETROGRADE | SYLLEPSIS | ZIGZAG) #IMPLIED>
<!ATTLIST NARRATOR distance (NARRATED | INDIRECT | FREE_INDIRECT | REPORTED | DIRECT | IMMEDIATE) #IMPLIED>
<!ATTLIST NARRATOR perspective (NON_FOCALIZED | INTERNALLY_FOCALIZED | EXTERNALLY_FOCALIZED | OTHER) #IMPLIED>
<!ATTLIST NARRATOR timeRelation (SUBSEQUENT | SIMULTANEOUS | PRIOR) #IMPLIED>
<!ATTLIST NARRATOR speechTime IDREF #IMPLIED>
<!ATTLIST NARRATOR person CDATA #IMPLIED>
<!ELEMENT AUDIENCE EMPTY>
<!ATTLIST AUDIENCE id ID #REQUIRED>
<!ATTLIST AUDIENCE name CDATA #IMPLIED>
<!ATTLIST AUDIENCE form CDATA #IMPLIED>
<!ELEMENT CHARACTER (#PCDATA)>
<!ATTLIST CHARACTER id ID #REQUIRED>
<!ATTLIST CHARACTER exists (true | false) #IMPLIED>
<!ATTLIST CHARACTER accessibleTo IDREFS #IMPLIED>
<!ATTLIST CHARACTER name CDATA #IMPLIED>
<!ATTLIST CHARACTER form CDATA #IMPLIED>
<!ATTLIST CHARACTER type (animate | inanimate) #IMPLIED>
<!ATTLIST CHARACTER attributes CDATA #IMPLIED>
<!ELEMENT EVENT (#PCDATA)>
<!ATTLIST EVENT id ID #REQUIRED>
<!ATTLIST EVENT exists (true | false) #IMPLIED>
<!ATTLIST EVENT type (ACTION | SPEECHACT | PERCEPTION | MENTAL) #IMPLIED>
<!ATTLIST EVENT participants IDREFS #IMPLIED>
<!ELEMENT TIME (#PCDATA)>
<!ATTLIST TIME id ID #REQUIRED>
<!ATTLIST TIME accessibleTo IDREFS #IMPLIED>
<!ATTLIST TIME value NMTOKEN #IMPLIED>
<!ELEMENT PLACE (#PCDATA)>
<!ATTLIST PLACE id ID #REQUIRED>
<!ATTLIST PLACE exists (true | false) #IMPLIED>
<!ATTLIST PLACE accessibleTo IDREFS #IMPLIED>
<!ELEMENT EVALUATION EMPTY>
<!ATTLIST EVALUATION id ID #REQUIRED>
<!ATTLIST EVALUATION eventID IDREF #REQUIRED>
<!ATTLIST EVALUATION characterID IDREF #REQUIRED>
<!ATTLIST EVALUATION audienceID IDREF #REQUIRED>
<!ATTLIST EVALUATION value (1 | 2) #IMPLIED>
<!ATTLIST EVALUATION polarity (positive | negative | neutral) #IMPLIED>
]>
```

Fig. 3.1 NarrativeML: initial version

in all, this is a barebones version of NarrativeML based on the distinctions introduced so far; it will be elaborated over subsequent chapters.[9]

LLMs Generating NarrativeML

Example 3.22 shows a story from the ROC data, along with the NarrativeML auto-generated for it by GPT-4o.

---

[9] This NarrativeML DTD leaves out character offsets for mentions of particular tags in the text. While the final DTD version, in Chap. 7, has it, we leave out character offsets from all examples in this book as they are cumbersome to read. In practical use, we avail of standoff rather than inline annotations, as discussed in the Appendix, but for pedagogical purposes, inline annotations are often shown.

3.6 NarrativeML

Example 3.22

User Prompt: Here is the curated NarrativeML for this sentence: *March 7, 2006. Leaving San Cristobal de las Casas, I biked with Gregg and Brooks for one more day.* and its DTD ...
Now, generate the NarrativeML for this text:
Yesterday, I went to my favorite place to buy groceries. I decided to go in the afternoon in my car. I usually buy the same things, but today I bought some fresh steaks. I couldn't wait to get home and grill them for dinner in the evening. Everyone enjoyed the dinner and it was another great dinner.
Machine:
```
<NarrativeML version="0.4">
  <NARRATIVE id="n1" narrator="n2" title="Grocery Trip and Dinner"
  medium="first-person_narrative">
    <NARRATOR id="n2" name="Narrator" type="present"
    order="CHRONICLE" distance="NARRATED"
    perspective="INTERNALLY_FOCALIZED"/>
    <CHARACTER id="c1" name="Narrator" type="animate"/>
    <CHARACTER id="c2" name="Everyone" type="animate"/>
    <PLACE id="p1">Favorite grocery store</PLACE>
    <PLACE id="p2">Home</PLACE>
    <SEGMENT id="s1" title="Grocery Trip and Dinner">
      <TIME id="t1" value="2024-10-18">Yesterday</TIME>
      <TIME id="t2" value="Afternoon">afternoon</TIME>
      <TIME id="t3" value="Evening">evening</TIME>
      <EVENT id="e1" type="ACTION" participants="c1 p1"
      duration="PT30M">I went to my favorite place to buy
      groceries</EVENT>
      <EVENT id="e2" type="ACTION" participants="c1 p1"
      duration="PT10M">I decided to go in my car</EVENT>
      <EVENT id="e3" type="ACTION" participants="c1 p1"
      duration="PT20M">I bought fresh steaks</EVENT>
      <EVENT id="e4" type="MENTAL" participants="c1"
      duration="PT5M">I couldn't wait to get home and grill
      them for dinner</EVENT>
      <EVENT id="e5" type="ACTION" participants="c1 p2"
      duration="PT1H">Grilled steaks for dinner</EVENT>
      <EVENT id="e6" type="ACTION" participants="c1 c2 p2"
      duration="PT1H">Everyone enjoyed the dinner</EVENT>
      <EVALUATION id="eval1" eventID="e6" characterID="c1"
      audienceID="c2" value="1" polarity="positive">It was another
      great dinner</EVALUATION>
        </SEGMENT>
  </NARRATIVE>
</NarrativeML>
```

GPT-4o mistakenly tags the evaluative sentence "It was another great dinner" as an EVALUATION, whereas the latter is reserved for audience evaluation. 'Afternoon' and 'Evening' aren't appropriate ISO values for Times—they should be 'XXXX-XX-XXTAF' and 'XXXX-XX-XXTEV', respectively (Ferro et al. 2005). The extents of events being tagged are also not delimited properly. But otherwise, it seems reasonable, even adding in helpful comments. Note that I deliberately chose to give the LLM the barebones DTD from Fig. 3.1, and a rudimentary annotated example. As we shall see later in the book, given the final DTD fleshed out by the end of Chap. 6, and given a few more examples, the output becomes far more accurate.

NarrativeML is not intended to subsume or supplant other annotation schemes. We will assume that the predicates and arguments for each sentence are available under separate annotation based on the PropBank scheme of Palmer et al. (2005). That would make the agent initiating the leaving being the same as the one doing the biking, which will coincide in turn with the narrator. As with the PropBank annotation scheme, we will assume that factuality profiles are part of an addition layer of annotation, external to NarrativeML. Likewise, other annotation layers such as (Pustejovsky et al. 2005), SpatialML (Mani et al. 2010), and ISO-Space (Mani et al. 2012) could also be added. We will investigate imports from TimeML in Chap. 5.

3.7 Conclusion

The chapter began by tracing narratology's origins as a foundational field. We defined Narrator Identity as the answer to "Who speaks?" and observed that LLMs can often determine this through anaphoric and coreference resolution. Next, we discussed Narrative Distance, explaining concepts like mimesis and diegesis and exploring a continuum from narrated to transposed speech, culminating in various quotation forms. Examining focalization, or "Who sees?", we found the traditional three-way distinction (external, internal, zero-focalization) to be oversimplified. Using the ROC corpus, we conducted focalization exercises, varying focalizations and assessing plausibility, noting that LLMs are able to implicitly detect certain types of focalization and to detect it explicitly when instructed to, and moreover to alter the focalization. However, they often struggle with mixed focalization cases.

We then addressed embedded narratives, which we found to be manageable for LLMs when structurally simple, but challenging when in complex forms. Similarly, disentangling narrative threads, which are increasingly prevalent in modern literary methods, poses difficulties for LLMs.

Turning temporarily away from LLMs per se, subordinated discourse, essential in both narration and mental events of characters, emerged as a crucial but under-explored aspect in traditional narratology. This led to a digression on the importance of modal reasoning, noting that though computationally challenging, machine learning can help approximate by indicating the scopes of intensional predicates. We discussed representing accessibility relations across complex narratives. We also emphasized character representation and goals, introducing a new theory of audience satisfaction through character evaluations involving emotions.

Bringing these elements together, we introduced NarrativeML, a practical annotation scheme. Despite its complexity, NarrativeML annotations can be effectively handled by LLMs, showing its expressive power as further expanded in later chapters.

For further discussion of narratology and illuminating how it broadly relates to certain computational approaches, see the excellent overview by Piper et al. (2021).

References

Aristotle (1932). *Poetics*. Project Gutenberg, Number 1974.
Bamman, David, Ted Underwood, and Noah A. Smith (2014). "A Bayesian Mixed Effects Model of Literary Character". In: *Proceedings of the 52nd Annual Meeting of the Association for Computational Linguistics*, pp. 370–379. https://aclanthology.org/P14-1035/.
Booth, Wayne C. (1983). *The Rhetoric of Fiction*. Chicago: University of Chicago Press.
Byatt, A. S. (1994). "The Djinn in the Nightingale's Eye". In: *The Paris Review*. issue 33, Winter 1994.
Cortazar, Julio (1967). *End of the Game and Other Stories*. New York: Pantheon Books. https://biblioklept.org/2023/06/09/a-continuity-of-parks-julio-cortazar/.
Edmiston, William F. (1991). *Hindsight and Insight: Focalization in Four Eighteenth-Century French Novels*. University Park, PA: Penn State University Press.
Elson, David K. (2012). "Modeling Narrative Discourse". PhD thesis. Department of Computer Science, Columbia University.
Elson, David K., Nicholas Dames, and Kathleen R. McKeown (2010). "Extracting social networks from literary fiction." In: *Proceedings of the 48th Annual Meeting of the Association for Computational Linguistics (ACL'2010)*. Uppsala, Sweden, pp. 138–147. https://aclanthology.org/P10-1015/.
Ferro, Lisa, Laurie Gerber, Inderjeet Mani, Beth Sundheim, and George Wilson (2005). *TIDES 2005 standard for the annotation of temporal expressions*. Tech. rep. McLean, VA: MITRE. http://timex2.mitre.org/annotation%5C_guidelines/timex2%5C_annotation%5C_guidelines.html.
Fitzgerald, Robert (1975). *The Iliad of Homer*. New York: Anchor Press/Doubleday.
Forster, E. M. (1956). *Aspects of the Novel*. New York: Harcourt.
Genette, Gerard (1980). *Narrative Discourse*. Ithaca: Cornell University Press.
Genette, Gerard (1988). *Narrative Discourse Revisited*. Ithaca: Cornell University Press.
Halpern, Joseph Y. and Yoram Moses (1992). "A guide to completeness and complexity for modal logics of knowledge and belief." In: *Artificial Intelligence* 54, pp. 319–379. http://dx.doi.org/10.1016/0004-3702(92)%2090049-4.
Herman, David (2014). "Cognitive narratology". In: *Handbook of Narratology*. Ed. by Peter Huhn, Jan Christoph Meister, John Pier, and Wolf Schmid. Berlin: De Gruyter, pp. 46–64.
Hobbs, Jerry R. (1985). "Ontological promiscuity." In: *Proceedings of the 23rd Annual Meeting of the Association for Computational Linguistics (ACL)*. Chicago, Illinois, pp. 61–69. http://dx.doi.org/10.3115/981210.981218.
Hogan, Patrick (2003). *The Mind and Its Stories: Narrative Universals and Human Emotion*. New York: Cambridge University Press.
Iser, Wolfgang (1974). *The Implied Reader: Patterns of Communication in Prose Fiction from Bunyan to Beckett*. Baltimore: Johns Hopkins University Press.

Iyyer, Mohit, Anupam Guha, Snigdha Chaturvedi, Jordan Boyd-Graber, and Hal Daume III (2016). "Feuding Families and Former Friends: Unsupervised Learning for Dynamic Fictional Relationships". In: *Proceedings of the North American Association for Computational Linguistics (NAACL'2016)*. https://home.cs.colorado.edu/~jbg/docs/2016_naacl_relationships.pdf.

Jin, Jiyan and Siyun Liu (2024). "The influences of narrative perspective shift and scene detail on narrative semantic processing". In: *Language and Cognition, First View*, pp. 1–29. https://doi.org/10.1017/langcog.2024.9.

Kripke, Saul (1959). "A completeness theorem in modal logic". In: *Journal of Symbolic Logic* 24.1, pp. 1–14. http://dx.doi.org/10.2307/2964568.

Mani, Inderjeet (2014). "Computational narratology". In: *Handbook of Narratology*. Ed. by Peter Huhn, Jan Christoph Meister, John Pier, and Wolf Schmid. Berlin: De Gruyter, pp. 84–92.

Mani, Inderjeet and James Pustejovsky (2012). *Interpreting Motion: Grounded Representations for Spatial Language*. New York: Oxford University Press.

Mani, Inderjeet, Christine Doran, et al. (2010). "SpatialML: annotation scheme, resources and evaluation". In: *Language Resources and Evaluation* 44.3, pp. 263–280. http://dx.doi.org/10.1007/s10579-010-9121-0.

McHale, Brian (2003). *Postmodernist Fiction*. London: Routledge.

Meister, Jan Christoph (2014). "Narratology". In: *Handbook of Narratology*. Ed. by Peter Huhn, Jan Christoph Meister, John Pier, and Wolf Schmid. Berlin: De Gruyter, pp. 623–645.

Montfort, Nick (2011). "Curveship's automatic narrative variation". In: *Proceedings of the 6th International Conference on the Foundations of Digital Games (FDG '11)*. Bordeaux, France, pp. 211–218.

Moretti, Franco (1999). *Atlas of the European Novel, 1800-1900*. London: Verso.

Mostafazadeh, Nasrin, Nathanael Chambers, et al. (2016). "A corpus and evaluation framework for deeper understanding of commonsense stories". In: *Proceedings of the North American Chapter of the Association for Computational Linguistics: Human Language Technologies (NAACL-HLT'2016)*, pp. 839–849. https://aclanthology.org/N16-1098.pdf.

Palmer, Martha, Dan Gildea, and Paul Kingsbury (2005). "The Proposition Bank: a corpus annotated with semantic roles." In: *Computational Linguistics* 31.1, pp. 71–105. http://dx.doi.org/10.1162/0891201053630264.

Pang, Bo, Lillian Lee, and Shivakumar Vaithyanathan (2002). "Thumbs up? sentiment classification using machine learning techniques." In: *Proceedings of the Conference on Empirical Methods in Natural Language Processing (EMNLP'2002)*. Philadelphia, PA, pp. 79–86. http://dx.doi.org/10.3115/1118693.1118704.

Parsons, Terence (1980). *Nonexistent Objects*. New Haven: Yale University Press.

Phelan, James (1989). *Reading People, Reading Plots: Character, Progression and the Interpretation of Narrative*. Chicago: University of Chicago Press.

Pier, John (2014). "Metalepsis". In: *Handbook of Narratology*. Ed. by Peter Huhn, Jan Christoph Meister, John Pier, and Wolf Schmid. Berlin: De Gruyter, pp. 326–343.

Piper, Andrew, Richard Jean So, and David Bamman (2021). "Narrative Theory for Computational Narrative Understanding". In: *Proceedings of the 2021 Conference on Empirical Methods in Natural Language Processing*, pp. 298–311. https://aclanthology.org/2021.emnlp-main.26/.

Plato (1871). *Republic*. Project Gutenberg, Number 1497.

Prince, Gerald (1982). *Narratology: The Form and Functioning of Narrative*. Berlin: Mouton.

Pustejovsky, James, Bob Ingria, et al. (2005). "The specification language TimeML". In: *The Language of Time: A Reader*. Ed. by Inderjeet Mani, James Pustejovsky, and Robert Gaizauskas. New York: Oxford University Press, pp. 549–562.

Quinn, Arthur (1982). *Figures of Speech*. Layton, Utah: Gibbs Smith.

Rabinowitz, Peter (1988). *Before Reading: Narrative Conventions and the Politics of Interpretation.* Columbus: Ohio State University Press.

Sauri, Roser and James Pustejovsky (2008). "From structure to interpretation: a double-layered annotation for event factuality". In: *Proceedings of the Second Linguistic Annotation Workshop.* Marrakech, Morocco. https://www.cs.brandeis.edu/~roser/pubs/sauriPustejovsky_lrec08_2.pdf.

Sauri, Roser and James Pustejovsky (2012). "Are you sure that this happened? Assessing the factuality degree of events in text". In: *Computational Linguistics* 38.2, pp. 1–39. http://dx.doi.org/10.1162/COLI%5C_a%5C_00096.

Scholes, Robert and Robert Kellogg (1996). *The Nature of Narrative.* Oxford: Oxford University Press.

Shklovsky, Viktor (1973). "On the connection between devices of syuzhet construction and general stylistic devices". In: *Russian Formalism.* Ed. by S. Bann and J. E. Bowlt. Edinburgh: Scottish Academic Press. http://dx.doi.org/10.1093/jts/flr173.

Todorov, Tzvetan (1981). *Introduction to Poetics.* Minneapolis: University of Minnesota Press.

Wallace, Byron C. (2012). "Multiple narrative disentanglement: unraveling Infinite Jest". In: *Proceedings of the North American Chapter of the Association for Computational Linguistics: Human Language Technologies (NAACL-HLT'2012).* Montreal, Canada, pp. 1–10. https://aclanthology.org/N12-1001/.

Plot Basics

4.1 Introduction

In this chapter, I will introduce classical narratological notions of plot, beginning with widely accepted causal definitions. I explain the classical Aristotelian notions of plot, the narrative arc of Freytag, and the Heroic Quest of Campbell (both of which are popular among scriptwriters). Next comes a discussion of Propp's narrative functions, a topic that has stimulated computational interest, before we turn to models of plot from the AI community, all of which are graph-based, including Plot Units, Doxastic Preferences, Story Intention Graphs, and Narrative Event Chains. The chapter clarifies how LLMs and supplementary methods including common sense knowledge graphs can be used to arrive at some of these representations. While much of the chapter focuses on analysis, examples are also given of generation and summarization. I then compare these different models and suggest some of the ramifications for humanities narratology. I then add a light plot-related layer to NarrativeML, that will be elaborated further in subsequent chapters.

4.1.1 Background

Events are a key aspect of narrative. The *causal* connection between them has often been viewed as a factor influencing what makes a narrative compelling. The term **plot** is used to represent this causal aspect. As Forster (1956) argues, while what he calls a *story* is "a narrative of events arranged in their time-sequence," a *plot* is "also a narrative of events, the emphasis falling on causality." To quote from Forster (1956, p. 86):

> "The king died and then the queen died," is a story. "The king died, and then the queen died of grief" is a plot. The time-sequence is preserved, but the sense of causality overshadows it. Or again: "The queen died, no one knew why, until it was discovered that it was through grief at

the death of the king." This is a plot with a mystery in it, a form capable of high development. It suspends the time-sequence, it moves as far away from the story as its limitations will allow. Consider the death of the queen. If it is in a story we say: "And then?" If it is in a plot we ask: "Why?".

Plot in literary theory thus focuses on causal explanations for the events in the narrative. These can unfold due to the actions of agents or of other event participants, or other forces in the world.

Our everyday use of the term *plot* is somewhat looser, and covers summaries of what happened in the story (spoiler or not). It can involve a partial explanation of why things happened, and in that case is sometimes grounded in terms of inferences about agents' goals and beliefs. How much drill-down this involves can depend on the genre as well as who the reader or audience is. Clearly, certain types of events can recur across stories, e.g., marriage, death, abduction, or rescue; and some of the more interesting stories involve complex sequences of events such as punishment, betrayal, retaliation, etc.

Some informal plots provide only an abstract summary of the fabula, e.g., *X met Y at the prom; They dated; They split up; They got married.* This involves classifying event types in the fabula, leaving it to the audience to use background knowledge to guess in most cases why X might have done those things. We call these types of plot models **event summaries**.

Many modern modern works of so-called literary fiction (as opposed to genres such as detective or romance or fantasy stories) are less constrained by plot; here one might count novels such as Woolf's *The Waves* or Calvino's *Mr. Palomar*. At the same time, many more works, including the *Iliad* and the *Mahabharata*, have extremely complex plots that the reader will not be able to keep track of. Some critics also try to classify all stories into being either *character-driven* or *plot-driven*. The point is, however, that even if a work is extremely difficult, or written so as to disregard various conventions for 'plotting', humans have a tendency to find structure and meaning in almost any arrangement of events.

4.1.2 Aristotelian Plot

The emphasis on the causal connection between events in a narrative goes back to Aristotle (1932) in the 4th century BCE. He defines *mythos*, which is often translated as 'plot', as a sequence of events linked by necessity or probability. Focusing on the genres of <u>epic poetry</u> and <u>drama</u>, he distinguishes plot from other narrative sequences of events such as biography, where incidents in a person's life may lack a necessary and probable connection, and history, where (he says) events are confined to what actually happened, rather than hypothetical happenings. Plot is important, because the spectator reconstructs it in responding emotionally to the language of the drama: "For the plot ought to be so constructed that, even without the aid of the eye, he who hears the tale told will thrill with horror and melt to pity at what takes place." (Aristotle 1932) (XIV).

4.1 Introduction

In applying these notions to epic poems and tragic dramas, Aristotle defends his emphasis on events rather than characters. "Character determines men's qualities, but it is their actions that make them happy or wretched" (ibid., VI). However, plot is just one aspect of tragedy; the other aspects he included were character, diction, thought, spectacle, and song.

Aristotle's conception of plot involves *aggregation* over entire sequences of events, as in his plot for the *Odyssey*, from Aristotle (1932) (XVII):

Example 4.1
A certain man is absent from home for many years; he is jealously watched by Poseidon, and left desolate. Meanwhile his home is in a wretched plight—suitors are wasting his substance and plotting against his son. At length, tempest-tost, he himself arrives; he makes certain persons acquainted with him; he attacks the suitors with his own hand, and is himself preserved while he destroys them.

Not all events are party to the plot: incidental events (which Aristotle calls *episodes*), while obviously present in a drama, do not form part of the plot "for a thing whose presence or absence makes no visible difference, is not an organic part of the whole."

In essence, a plot, like the forms of other imitative arts, has a *unity of action*, involving an action that is complete and whole, and with a property of *compactness*, "the structural union of the parts being such that if any one of them is displaced or removed, the whole will be disjointed and disturbed."

Implausible plots, in Aristotle's definition, can involve unlikely or poorly motivated happenings. There are countless examples of these, some even flowing from the pens of gifted writers. For example, in Faulkner's novel *Wild Palms*, an impecunious medical intern just happens to discover, in a trash can, a wallet full of cash, which allows him to elope with a sculptress, their mutual attraction lacking any convincing prior development.

Here it is worth noting that there has been little practical attention paid to how long a plot should be. In online corpora of book summaries such as the CMU Book Summary Dataset (DavidSmith 2013), we see that when novels are present, the so-called plot summaries tend to include key events, themes, characters, and sometimes even fragments of dialogue. Many of these summaries are overly long (though on the average only 1% of the book length) and too review-like to be considered as plots.

4.1.3 Narrative Arc

The Aristotelian analysis develops a basic form of what is called the *narrative arc*. In more detail, from Aristotle (1932):

- He specifies a number of prescriptive guidelines for constructing suitable plots. These include having scenes of recognition, or *anagnorisis*, what he calls a change from ignorance to knowledge: "in the recognition of Odysseus by his scar, the discovery is made in one way by the nurse, in another by the swineherds." He also favors reversals of fortune, or *peripeteia* (ibid., XI):

> "Thus in *Oedipus*, the messenger comes to cheer Oedipus and free him from his alarms about his mother, but by revealing who he is, he produces the opposite effect. ... The change of fortune should be ... from good to bad."

- He provides a three-phase substructure for a plot, with a beginning, middle, and end (ibid., VII):

> "A beginning is that which is not a necessary consequent of anything else but after which something else exists or happens as a natural result. An end on the contrary is that which is inevitably or, as a rule, the natural result of something else but from which nothing else follows; a middle follows something else and something follows from it."

- Aristotle identifies a crucial inflexion point in the middle, called a turning point, which is a *complication* followed by an *unraveling* (ibid., XVIII):

> "By the complication I mean all that extends from the beginning of the action to the part which marks the turning-point to good or bad fortune. The unraveling is that which extends from the beginning of the change to the end."

- In identifying beginning, middles, and ends, Aristotle seems to suggest that these are in order, i.e., an ordering of the discourse, not just the fabula. This is somewhat puzzling, since Aristotle was familiar with epic poetry as well as tragedies that began in the mode which Horace (1976) later (in the first century BCE) called *in medias res*, i.e., in the middle of things. Certainly, epic poetry, when recited orally, could vary the order of events in each telling.

Aristotle's three-part decomposition has been further refined by many others, especially Freytag (1900). In his scheme, drama goes through five successive stages, in keeping with the classical five-act play, illustrated by Abrams (1993) from *Hamlet*:

1. Introduction, also called Exposition. This is the setting and prior background, involving the sentinels meeting the ghost, "the starting call of the watch, the mounting of the guard, the appearance of the ghost" (Freytag, p. 118).
2. Rise, also called Rising action. This starts when Hamlet meets his father's ghost (Act 1, Scene 4), who reveals that the father was murdered by Claudius. This continues with Hamlet controlling, more or less, the course of events.
3. Climax. This begins (Act 3, Scene 2) when Hamlet confirms that Claudius is guilty when the latter leaves the room during the murder scene of the embedded play *The Murder of Gonzago*. It corresponds to Aristotle's notion of *complication*.

4.1 Introduction

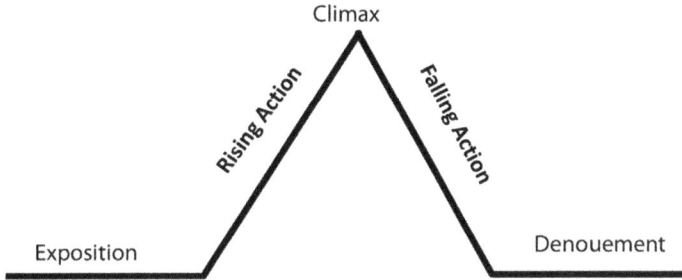

Fig. 4.1 Freytag's analysis of drama

4. Fall, also called Falling action. This begins (Act 3, Scene 3) with Hamlet's failure to kill the king at prayer. The course of events is more or less controlled by Claudius.
5. Catastrophe, also called denouement. This is the catastrophe (Act 5, Scene 2) that ends with the deaths of Hamlet, Claudius, and the Queen. This corresponds to Aristotle's *unraveling*.

Freytag's narrative arc is often pictured as in Fig. 4.1.

Freytag's analysis, unlike Aristotle's, does not focus specifically on events, or sequences thereof. It is often applied to highly varied units of discourse, from short passages to entire scenes of a play. When applied to the Greek and Shakespearean drama for which it was intended, it provides many examples but stops short of offering precise criteria as to what counts as an instance of each of these stages. The Freytagian scheme also does not apply to epic poetry, where the narrated order of events can vary substantially with each oral performance.

The narrative arc has nevertheless been relatively popular as a rubric for dramatic writing that focuses on plot. It has influenced screenwriting, where many formulas that are taught include a three-act decomposition that blends Aristotle and Freytag. It has also influenced computational narratology, in particular, interactive storytelling, e.g., via extensions proposed by Laurel (1993), Murray (1997), Mateas (2000), and Mateas and Stern (2005). The key idea is to impose structure on an overall storytelling experience by forcing the narrative to follow a pre-specified narrative arc.

4.1.4 Heroic Quests

Another popular rubric for plot is that of the **heroic quest**, where the inspiration comes from mythology. The study of comparative religion and mythology by Campbell (1949), described in his 1949 best-selling book *The Hero with a Thousand Faces*, was carried out within a tradition of Jungian psychoanalytic theory. Campbell found a common pattern across some of the world's myths, reflective of a rite of passage, that he called *monomyth* (a word borrowed from James Joyce's *Finnegan's Wake*). It involves the following scenario:

Example 4.2
A hero ventures forth from the world of common day into a region of supernatural wonder: fabulous forces are there encountered and a decisive victory is won: the hero comes back from this mysterious adventure with the power to bestow boons on his fellow man.

There are three stages to the monomyth, which bottom-out in 17 stages in all:

1. Departure

    a. The Call to Adventure
    b. Refusal of the Call
    c. Supernatural Aid
    d. The Crossing of the First Threshold
    e. Belly of The Whale

2. Initiation

    a. The Road of Trials
    b. The Meeting With the Goddess
    c. Woman as Temptress
    d. Atonement with the Father
    e. Apotheosis
    f. The Ultimate Boon

3. Return

    a. Refusal of the Return
    b. The Magic Flight
    c. Rescue from Without
    d. The Crossing of the Return Threshold
    e. Master of Two Worlds
    f. Freedom to Live

Prototypical examples of this monomyth are found in the story of Jason and the Golden Fleece, Aeneas and the founding of Rome, the life of the Buddha, the story of Moses, and so forth. Campbell's ideas, focused as they are on the most ancient stories across cultures (as well as by extensive reading of Joyce), have influenced critics as well as motivated writers,

4.1 Introduction

filmmakers, musicians, game designers, storytelling system developers, etc. Given the wide influence of these ideas, it is worthwhile examining them carefully.

In prescribing as it does an overall master plot, the monomyth tries to emphasize what Campbell believes are the universal commonalities across stories, without focusing on individual differences that arise in each particular instance.

As Dundes (2005) points out, the monomyth account has several shortcomings:

- Campbell is apparently unaware of the folkloric distinction between myths (which are religious explanations, as in creation myths), legends, which are narrations of historical or pseudo-historical events (e.g., the stories of Robin Hood), and folk tales, which are stories such as fairy tales, that are intended to be fictions.
- His similarity judgments are also far-fetched: the 'Belly of the Whale' notion tends to conflate the story of Jonah (male swallowed by whale) with that of Red Riding Hood (female devoured by wolf, but only in some versions).
- His claims of universality for certain themes (e.g., virgin birth) ignore the actual and skewed geographical distribution of such themes.

While the heroic quest itself is a narrative motif well-known to scholars, its characterization by Campbell has traveled far beyond that. His account has directly impacted screenwriting, for example, in Dan Harmon's Story Circle,[1] and also helped in the design and analysis of role-playing games. Despite its fame, the heroic quest is clearly a genre-specific, coarse-grained framework whose mapping to the fabula is unspecified. It is too abstract to provide enough of a detailed description that can be used to track plot given linguistic input, or to formulate a plot in such a way that it has a bearing on the events in the fabula and their expression in the discourse. There are no specifications of required versus optional stages, or of dependencies within stages. It is also not really of much use for even the most basic of tales, such as Aesop's *The Wily Lion*, described a little later in Sect. 4.1.6.4.

4.1.5 Narrative Functions

Campbell's scheme was heavily influenced by Vladimir Propp's set of 31 ordered narrative elements, called *functions* or *narratemes*, first published in Propp's seminal book *Morphology of the Folktale* in Russian in 1928, see Propp (1968). Propp's work is attractive because of its formal nature, backed by extremely detailed analyses of 10 sample stories drawn from a corpus of 100 Russian folk tales. To motivate his account, he considers stories which at the level of **narrative functions** seem very similar, even though the characters involved are different (from Propp 1968, p. 8):

---

[1] https://en.wikipedia.org/wiki/Dan_Harmon.

1. "A tsar gives an eagle to a hero. The eagle carries the hero away to another kingdom."
2. "An old man gives Sucenko a horse. The horse carries Sucenko away to another kingdom."
3. "A sorcerer gives Ivan a little boat. The boat takes Ivan to another kingdom."
4. "A princess gives Ivan a ring. Young men appearing from out of the ring carry Ivan away into another kingdom."

In the above cases, all involve a common narrative function that Propp calls 'Receipt of a Magical Agent'. Narrative functions have four characteristics (ibid., p. 10):

1. "Functions of characters (*narratemes*) serve as stable, constant elements in a tale, independent of how and by whom they are fulfilled."
1. "A function is an act of a character, defined from the point of view of its significance for the course of the action."
2. "The number of functions known to the fairy tale is limited."
3. "The sequence of functions is always identical."

The complete set of 31 narrative functions is shown in Table 4.1 (from Propp 1968).

Table 4.1 Propp's narrative functions

Number	Symbol	Function	Number	Function	Symbol
1	β	Absentation	16	Struggle	H
2	γ	Interdiction	17	Branding	J
3	δ	Violation of interdiction	18	Victory	I
4	ϵ	Reconnaissance	19	Liquidation	K
5	ζ	Delivery	20	Return	$\downarrow$
6	η	Trickery	21	Pursuit	Pr
7	θ	Complicity	22	Rescue	Rs
8	A	Villainy/Lack	23	Unrecognized arrival	o
9	B	Mediation	24	Unfounded claims	L
10	C	Beginning counter-action	25	Difficult task	M
11	$\uparrow$	Departure	26	Solution	N
12	D	First function of the donor	27	Recognition	Q
13	E	Hero's reaction	28	Exposure	$Ex.$
14	F	Receipt of a magical agent	29	Transfiguration	T
15	G	Guidance	30	Punishment	U
31	W	Wedding			

4.1 Introduction

Each of these functions in turn has different types of subfunctions, indicated in Propp's notation by numerical superscripts. For example, Villainy (A) can involve a variety of different subfunctions, including one where the villain abducts someone (A^1), and Receipt of a Magical Agent (F) can involve a subfunction where "Various characters place themselves at the disposal of the hero" (F^9).

Characters are also classified into eight categories: Hero, Villain, Donor, Dispatcher, False Hero, Helper, Princess, Princess's Father. Finally, there are also about 150 *Elements*, some of which correspond to narrative functions; a mapping has been carried out by Scheidel and Declerck (2010).

Consider an example story *The Magic Swan Geese*, story number 113 from the corpus analyzed by Propp. Example 4.3 shows a condensed version of it.

Example 4.3
A girl was left in charge of her younger brother by her parents, but she didn't keep an eye on him and he was snatched away by swan geese. She chased after them, passing an oven, an apple tree, and a river of milk, each of which offered her delicacies which she declined. She finally arrived at a hut where Baba Yaga was sitting with her brother. Baba Yaga asked her to stay and spin flax, and left. A mouse came out and in exchange for porridge, revealed that Baba Yaga had gone to prepare the bath house so she could be steamed and then eaten. The girl fled with her brother, but Baba Yaga sent the swan geese after her. This time, seeking help from the river, apple tree, and oven, she accepted their offerings and was able to hide from the swan geese. She reached home with her brother in time for her parents' return.

The full story has been analyzed into the following complex sequence of narrative functions by Scheidel and Declerck (2010):

Example 4.4
$\gamma^1 \beta^1 \delta^1 A^1 C \uparrow \{[DE^n eg. Fneg.]^3 d^7 E^7 F^9\} G^4 K^1 \downarrow [Pr^1 D^1 E^1 F^9 = Rs^4]^3$

It is hard to believe a narrative can be boiled down to a sequence as in Example 4.4 but it illustrates the formal nature of Propp's analysis, and its computational relevance. The unraveling of the sequence, based on looking them up in Table 4.1, is left as an exercise for the reader.

LLMs and Narrative Functions
Other 'parses' are clearly possible for *The Magic Swan Geese*, and GPT-4o suggests several of them. In addition, given the parse in Example 4.4, it is able to generate a rather different story:

Example 4.5

> **User Prompt**: Can you generate a story based on this sequence of Propp's narrative functions . . . ?
> **Machine**: Once upon a time, a young prince (γ—Interdiction) was warned by his father not to leave the castle grounds, as great danger lurked outside.
> But the prince (β—Absentation) ignored the warning and snuck out into the forest, eager for adventure.
> (δ- Violation of Interdiction) His disobedience led him into the clutches of a cunning villain, who kidnapped him and hid him away in a dark tower (A—Villainy).
> When the prince's disappearance was discovered, his loyal companion (C—Beginning of Counteraction) set off on a quest to rescue him.
> He traveled far and wide ($\uparrow$—Departure), eventually meeting a wise old hermit (D—First Function of the Donor).
> The hermit tested the companion's courage, and upon passing the test (E—Hero's Reaction), he was granted a magical sword (F—Receipt of a Magical Agent).
> Guided by the sword's power (G—Guidance), the companion stormed the villain's fortress, engaging in fierce battle (H—Struggle) and ultimately defeating the villain (I—Victory).
> He freed the prince (K—Liquidation) and returned home ($\downarrow$—Return), but not without being pursued by the villain's soldiers (Pr—Pursuit).
> With help from his magical sword, the companion successfully eluded them and completed his mission (Rs—Rescue).

Our discussion so far reveals that Proppian functions are very abstract and cover many possible stories, which is a desirable criterion for a plot model.

An interesting approach to automatically labeling Proppian morphologies from a corpus is found in the work of Finlayson (2016). Based on linguistic annotation (in terms of TimeML, PropBank, and positive and negative affect) of a small corpus of 15 folk-tales, the algorithm used a machine-learning method to merge narrative structures found in stories to discover Proppian functions and characters. These were then evaluated against human annotations of those morphologies. As an example, consider these two micro-stories in Examples 4.6 and 4.7:

Example 4.6
The old man and the maidservant met on the road. He chased her, but she ran away. She thought he was ugly.

4.1 Introduction

Example 4.7
The dragon stalked the princess and scared her. She fled into hiding. She decided he was evil.

The learning method does model merging, where first the chasing and stalking events are merged (using measures of semantic similarity based on the WordNet thesaurus), then the running away and fleeing, and finally the thinking and deciding events. The eventual result is a generic story of Pursuit, Flight, and Judgment that constitutes an abstract level of Plot. Some interesting discoveries were made, for example, finding that in the corpus, kidnapping, seizing and tormenting are the hallmarks of Proppian villainy.

Attempts to take Proppian annotation further so as to make it of practical use on a large scale have met with some obstacles. For example, the recent work of Gervas and Mendez (2024) demonstrates that LLMs obtain an F-measure of no more than 0.4 with few-shot prompting on the Russian folk tales analyzed by Propp, suggesting there is considerable room for improvement.

Evaluations of Proppian annotation can however be challenging. As Bod (2012) pointed out, in their experiments in hand-annotating some of the folk tales in Propp's corpus:

- "Proppian descriptions of some of the dramatis personae (e.g., the Hero) and functions (e.g., H, i.e., Struggle) are vague and require a large amount of interpretation."
- "The Proppian framework encourages the marking of minor events that do not naturally occur in summaries of the same folktales."

Given these problems, they conclude that further inter-annotator studies are not worthwhile. However, they have been proved somewhat wrong by subsequent work by Yarlott and Finlayson (2016), who developed a markup language called ProppML. Annotating 15 Russian folktales using ProppML, they found high agreement between annotators once they were given sufficient training and time; the highest agreement was excellent (Kappa > 0.9) for narrative functions.

How applicable are Proppian morphologies to folktales from other cultures? The folklorist Dundes (1963) found they could be applied to Native American folktales, but with serious modifications of the patterns. The inter-disciplinary scholar Ramanujan (1991) found they could be applied to folktales from India, but again with local adaptations. He (1997) also found that Propp's morphology didn't work well for female-centered tales from Southern India. Studies have also been carried out of folk tales from China, Japan, the Philippines, and elsewhere. If the functions were general enough, one would expect a model trained or fine-tuned on ProppML to be able to predict the next turn in a folk tale or how it will end.

For now, however, there isn't enough annotated data to formally assess how general it is across cultures. One direction in which Propp's work can be taken to address this is to use the scheme introduced by Greimas (1984), who classifies characters into six categories: Subject, Object, Sender, Receiver, Helper, and Opponent, with sequenced relationships among them that are tied to desire, communication, and power. We will return to Greimas in Chap. 7, when we discuss conspiratorial narratives.

4.1.6 Causal Models of Plot

Narratives express facets of everyday experience, especially sets of events that need to be codified and shared in some way. This suggests immediately that representing fragments of narrative in memory, at an appropriate level of abstraction, can be useful for understanding and generating narratives. This idea dates back at least to Bartlett (1932), who asked subjects to repeat a given story over time. He found that the story got reconstructed in memory according to "an active organization of past reactions, or of past experiences" (ibid., p. 201), or **schema**. The schema imposes a level of abstraction on the information in a story, favoring gists over details, and biasing the assimilation of new information toward conformity with the schema, resulting in details irrelevant to the schema being left out.

4.1.6.1 Scripts
In the 1970s, AI research, particularly in the work at Yale of Roger Schank and his team (see Schank SchankAbelson 1977), gave these schemas a computational embodiment in terms of **scripts**, which are symbolic structures that represent a sequence of stereotypical events, such as eating at a restaurant. A restaurant script will involve characters typically found in a restaurant, such as waiters and patrons, and events such as arriving, perusing the menu (and perhaps other patrons), ordering, eating, paying a bill, and so forth, that may be arranged in a sequence. Characters have wants (such as food, i.e., they are hungry), and they can know that they can carry out actions (such as paying for the food). These are some of the preconditions that a script requires in order to be applicable. When a script gets 'executed', there are also postconditions that result, such as the person's hunger being satisfied. Scripts have to specify the characters, event sequences, preconditions, and postconditions.

As an example, the Script Applier Mechanism (SAM) program of Cullingford (1978) was an early story understanding system that came out of the Yale effort. It activated scripts based on the mentions, in the input story, of a script setting (e.g., a restaurant), precondition (e.g., a character being hungry), an event in the script, etc. Consider Example 4.8:

Example 4.8
John went to a restaurant. He ordered a hot dog. The waiter said they didn't have any. He asked for a hamburger. When the hamburger came, it was burnt. He left the restaurant.

SAM was able to answer commonsense inference questions such as whether John sat down in the restaurant (probably), whether he ordered a hot dog (yes), what the waiters served him (a hamburger), and whether John paid for his food (no, because he was angry over his burnt hamburger). Scripts are also viewed as essential in resolving discourse-dependent references. Thus, given the sentence "The hamburger came," SAM would use the restaurant script along with prior discourse information to infer that it was the waiter who brought the hamburger.

When someone enters a restaurant, it sets up expectations that the person is going to eat there (unless the person is a waiter, or a hired killer). If the person pays up, one can infer that she already ate (unless she was in a place where one pays first). A person may eat with the intent of not paying. Fast-food restaurants usually do not have waiters. Given all these different scenarios, a precise definition of what should be in a script becomes problematic. Even deciding how to distinguish scripts is a problem. As Schank and Abelson (1977) point out, in Example 4.9, it is unclear whether there is one script (taking a trip), two scripts (taking a trip and museum visit), or even three (bus trip, museum visit, and train trip).

Example 4.9
John took a bus to New York. In New York he went to a museum. Then he took a train home.

In order to work around these problems (Dejong 1982) introduced the notion of a *sketchy script*, which focuses only on the crucial events in a script, ignoring the rest. When processing a story, the mention of a key event could invoke an Arrest sketchy script. Once triggered, the other slots in the sketchy script, such as the crime, the culprit, etc., would be skimmed for, and hopefully, found in the text. The use of sketchy scripts reduces the knowledge required in a script, and allows for easier script selection. But if salient information that isn't coded in the script is present in the input, the script won't be triggered directly.

Research in the 1970s and 80s tried to address the problems with scripts by reorganizing and simplifying them as above, but they ended up with ad hoc specifications tied to the domains of interest. Thus, it became apparent that scripts were not in fact as generic as one would like from a computational standpoint. In addition, the disadvantages of the classic Aristotelian focus on events as opposed to characters (discussed in Chap. 3) became apparent. Soon it was obvious to the Yale group and others that characters in a story and their goals also needed to be modeled.

4.1.6.2 Plot Units

In the discussion so far, the models of plot have been behavioral, without taking into account the mental states of agents involved in events, and in particular ignoring the representation of their motives and goals. The account of plot offered by Lehnert (1981) represents characters' motivations for events, and their outcomes on characters (represented by positive ($+$) and negative ($-$) affect states, or emotions, for those characters). Motivations are represented as affect states (M) that are neutral. As events unfold, the causal relations between them are represented in terms of links between affect states of a given character involved in the event. A plot is viewed in terms of a graph of causal links between emotional states of characters; their beliefs and the events they are involved are not directly represented.

There are two types of causal links: a motivation link (m) between a negative outcome ($-$) and a mental state (M), reflecting the mental state caused by an external event, and an actualization link (a) between M and $+$, representing intentionality behind an event. In addition, there is a termination link (t) between mental states or between event outcomes, used when an event supplants or displaces the affective impact of a previous event, e.g., a second marriage *terminating* a divorce. Finally, there is an equivalence link (e) between mental states or between event outcomes, that represents a new situation that is similar to a previous situation. The termination and equivalence links thus run backward in time, in contrast to the (causal) motivation and actualization links, that run forward in time.

In addition to these four types of links involving a single character, there are cross-character links between affect states. For example, when, say, John wants to get his car started (M), and Paul agrees to help (M), there is a cross-character link between those two affect states, reflecting the fact that Paul has in this instance shared John's goal; and when Paul actualizes his goal to get it started ($+$), that results in a positive outcome for John ($+$), who has now actualized his own goal, albeit indirectly. Thus there is also a cross-character link between those latter positive affect states.

Particular recurring configurations in graphs are called **plot units**. The situation of Paul helping John by getting the latter's car started is a primitive plot unit called an Honored Request, shown in Fig. 4.2.

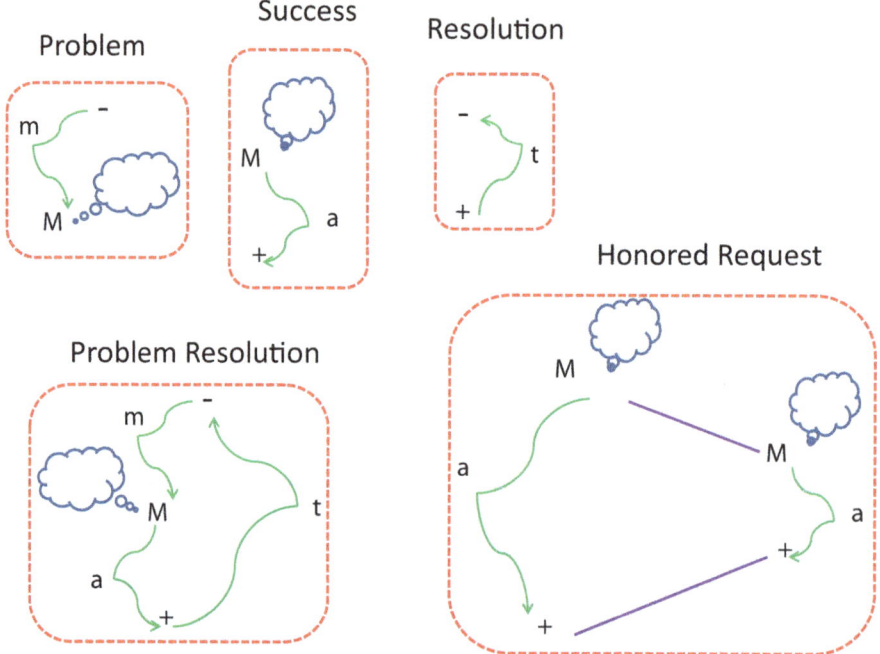

Fig. 4.2 Simple and complex plot units

4.1 Introduction

Figure 4.2 also reveals that an Honored Request is a complex plot unit made up of two primitive Success plot units, connected by cross-character links. Also shown is another complex plot unit called Problem Resolution, made up of three primitive plot units: Problem, Success, and Resolution.

Variants of an Honored Request include a Denied Request, where Paul does not necessarily share John's goal, being too busy; in this case, Paul's actualization, which by definition results in a positive outcome for him, results in (via a cross-character link) a negative outcome for John.

To see how primitive plot units are put together into complex plot units for a given narrative, consider Example 4.10 from Lehnert (1981):

Example 4.10
John was thrilled when Mary accepted his engagement ring. But when he found out about her father's illegal mail-order business, he felt torn between his love for Mary and his responsibility as a police officer. When John finally arrested her father, Mary called off their engagement.

The analysis of this example in terms of primitive and complex plot units is shown in Fig. 4.3, where the simple and complex plot units involved, Success ($S(M)$ and $S(J)$), Problem

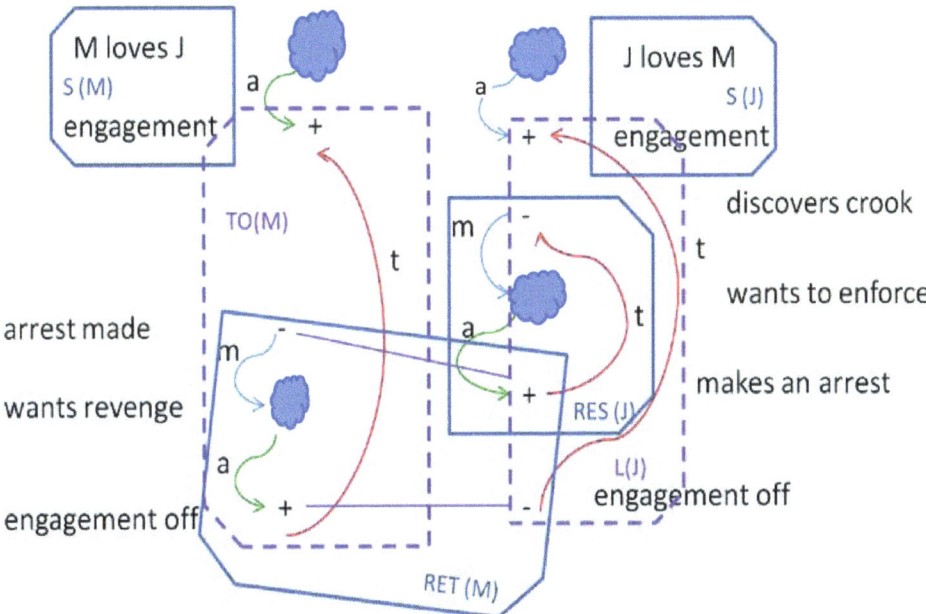

Fig. 4.3 Plot units in *Engagement* story

Resolution ($RES(J)$), Loss ($L(J)$), Retaliation ($RET(M)$), and Tradeoff ($TO(M)$) are indicated by labeled boxes.

Essentially, in the plot unit analysis, their being in love results in an actualization (a) of becoming engaged which results in a positive affect (+) for each. When he discovers Mary's father is a crook, John is in a negative affect state (−). This motivates (m) him to a mental state of wanting to enforce the law (M), which is actualized (a) by his making an arrest, resulting in a positive affect (+) for him, terminating (t) his negative affect. By a cross-character link, this results in a negative affect state (−) for Mary, who is motivated (m) to a mental state (M) of wanting revenge. This is actualized (a) by the event of her calling off the engagement, resulting in positive affect (+) for her, but by a cross-character link, a negative affect (−) for John, which results in terminating (t) his initial positive affect (+) from the engagement.

As can be seen, the plot unit analysis is both sophisticated and expressive. But is this plot a reasonable explanation of the causal structure of the narrative? Clearly, John's getting engaged is motivated by his being in love, which is a reasonable ground for doing so. John's arresting her father is based on his wanting to enforce the law; this does not, however, ground the explanation in terms of any higher moral goal or belief about the importance of laws.

Lehnert's analysis has a number of limitations:

- The model of emotion is severely limited (although such ternary classifications of emotion are hardly uncommon, having been prevalent since antiquity). Thus, in the Engagement story in Example 4.10, John's feeling torn between two affects is not directly expressible in the framework of Lehnert (1981). It is also awkward and moreover unrealistic to have all actualizations result in positive affect.
- Beliefs are not represented, restricting the grounding of explanations.
- The framework does not specify a complete set of primitive plot units, though Lehnert (1981) discusses at least 15.
- While the plot units themselves are defined precisely, there are no guidelines for whether to mark a mental state or event in a given situation, resulting in potential ambiguity. This in addition leaves the grounding of explanations rather open-ended.

At present, the reliability of annotators in annotating entire plots based on her scheme is not known, and neither has the plot unit approach been fully implemented.[2] However, Goyal et al. (2010) have developed, based on a corpus, a system that can infer, for each event, its possible affects (+, −, or M) for each human participant in the event. This is done by flagging verbs, especially those that involve mental states, speech events, etc., as well as those that impart a positive or negative polarity to their arguments (i.e., the verb's 'object', or *patient*), e.g., "rescued" versus "scammed." They next use the polarity information in affect projection rules to determine which participant (i.e., *patient* argument) receives which affect,

---

[2] The same is true of the enhanced account of Plot Units found in Ryan (1991), where agents' mental states are added to the representation.

4.1 Introduction

e.g., Mary in "Mary laughed," but the cat in "Mary hated the cat." They then compute actualization, termination, and motivation links between affect states using simple heuristic rules. The results for affect identification, over a test of 15 fables from Aesop, are a weak 0.45 F-measure. However, for link identification, their system scores 0.92 F-measure on actualization links and 0.72 F-measure on motivation links (with perfect, i.e., human, affect identification). The human agreement on affect identification is 0.7 Kappa, which is commonly viewed as high agreement. It turns out people largely agree on whether a verb is bad for the patient argument, but they do not necessarily agree if not whether it is good for the patient.

LLMs and Plot Units
How do LLMs fare on analyzing plot units?

Example 4.11

> **User Prompt**: Can you analyze this story in terms of plot units, affect states (positive and negative), and actualizations, including the causal chains? ...
> **Machine**: John was thrilled when Mary accepted his engagement ring (Success), achieving his goal of commitment and feeling positive affect: joy.
> His next goal was maintaining the relationship, but upon discovering her father's illegal business (Problem), he experienced negative affect: conflict between his love for Mary and his duty as a police officer. His new goal became resolving this moral dilemma, leading to the tradeoff: arresting her father to uphold the law, causing negative affect: guilt and anguish. As a consequence, Mary, feeling betrayed, called off the engagement (Loss), leaving John with negative affect: heartbreak and regret. Mary's retaliation, driven by her negative affect: betrayal and disappointment, severed the relationship (Retaliation). The causal chain of John's goals—from achieving love, to fulfilling duty, to losing both—triggers a shift from positive affect (joy) to negative affect (conflict, guilt, and loss) through key plot events.

GPT-4o's analysis of the Engagement story in Example 4.10 is reasonable, a little less technical than mine, but capturing the gist of the commonsense reasoning behind the story. This is yet another revelation of emergent knowledge in the LLM.

4.1.6.3 Doxastic Preferences

The research of Lowe (2010) describes a formalism called the Doxastic Preference Framework (DPF) for representing the expectations of characters in a narrative. It uses game trees

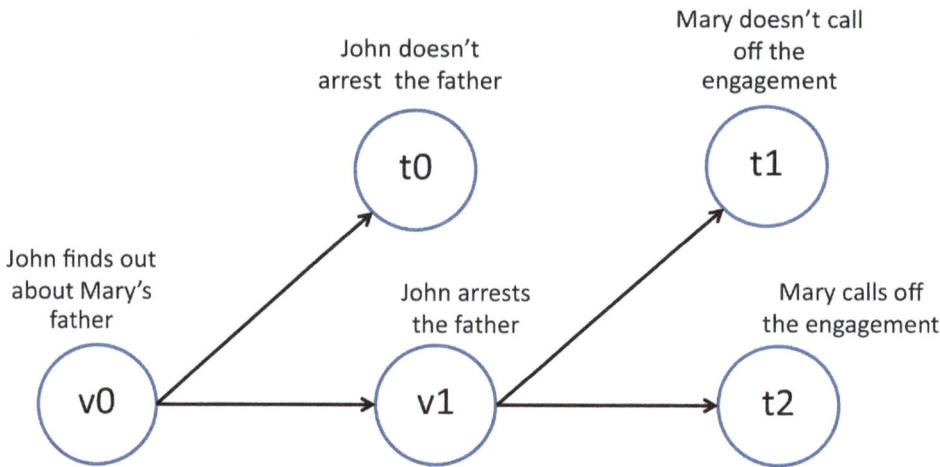

Fig. 4.4 *Engagement* story in DPF

that reflect the beliefs of characters about the preferences of other characters (as well as themselves) regarding the outcomes of events. The DPF analysis of the *Engagement* story from Example 4.10 is shown in the game tree Fig. 4.4.

We start in the non-terminal node $v0$, where John finds out about Mary's father. If he does not arrest him, the story would reach the terminal node $t0$. Instead, he does arrest him, and the story reaches non-terminal node $v1$. This results in an unexpected event for Mary (assuming no other events intervene, such as his sharing his discovery with her). If Mary decides not to call off the engagement, the story would end in terminal node $t1$. Instead, she ends the engagement, reaching terminal node $t2$. This is an unexpected action by Mary for John.

There are eight *building blocks*, such as Unexpected Action, Collaboration Gone Wrong, Betrayal, etc. These have been used to annotate a corpus of seven stories that are plot summaries of Crime Series Investigation episodes by Andel (2010). Unfortunately, no guidelines or inter-annotator reliability is available. In any case, for an annotator to hazard a guess as to what is a more likely reaction by one character to another's action can be problematic. While the scheme is quite expressive in terms of expressing beliefs about other's actions, and beliefs about beliefs, etc., it otherwise suffers from similar deficiencies to the scheme of Lehnert (1981).

4.1.6.4 Story Intention Graphs

More elaborated versions of plot that build on the idea of plot units are found in Elson (2012), who develops the concept of Story Intention Graphs (SIGs). SIGs are graph structures that include an interpretive layer linked to the propositions in the fabula as well as the text in the discourse. The interpretive layer consists of the beliefs, goals, and plans of agents, in

4.1 Introduction

addition to representing temporal relations between events. Consider a fable from Aesop, *The Wily Lion*:

Example 4.12
A Lion watched a fat Bull feeding in a meadow, and his mouth watered when he thought of the royal feast he would make, but he did not dare to attack him, for he was afraid of his sharp horns. Hunger, however, presently compelled him to do something: and as the use of force did not promise success, he determined to resort to artifice. Going up to the Bull in friendly fashion, he said to him, "I cannot help saying how much I admire your magnificent figure. What a fine head! What powerful shoulders and thighs! But, my dear friend, what in the world makes you wear those ugly horns? You must find them as awkward as they are unsightly. Believe me, you would do much better without them." The Bull was foolish enough to be persuaded by this flattery to have his horns cut off; and, having now lost his only means of defense, fell an easy prey to the Lion.

In annotating the interpretive layer for this story, the annotator has to represent implicit information such as the lion's belief that using force with the bull will result in his being gored by the lion, preventing his goal of eating the bull, and that the lion therefore has a goal to flatter the bull, so that the bull will trust the lion. The representation of such beliefs makes the SIG formalism more expressive than plot units, allowing for better-grounded explanations. Without strict guidelines, the choice of what goals to represent and which plans to use can be very open-ended.

A corpus of more than 60 stories, including fables from Aesop, stories by Chekhov and F. Scott Fitzgerald, etc., have been annotated by Elson (2012), creating a DramaBank. Here, the annotators are instructed only to annotate those goals and actions that are crucial to the evolution of the narrative. Inter-annotator studies by Elson (2012) have shown agreement of 0.55 Kappa, suggestive of moderate agreement. (Annotators were asked, in cases of plural or divergent interpretations, to settle on a single reading.) Deliberate authorial ambiguity, especially as found in literary texts, remains a problem. In addition, the goal-action representation is felt to be adequate for some texts, but too constraining for others. An annotator working on the SIG for a Chekhov story is quoted in Elson (2012, p. 197) as saying: *I think the main problem here is just that Chekhov's style is very difficult to break down into the goal-oriented structure, since large swaths of the story consist of characters looking over the water and realizing beauty and pointlessness.*

Analysis of common SIG patterns in the DramaBank show more than 80 patterns, giving rise to plot unit-like structures such as Deception, Change-of-Mind, Peripeteia, Anagnorisis, etc. The SIG approach departs from earlier work in specifying a mapping from semantic representations of sentences to the interpretive layer required for modeling the beliefs of characters. Here, unlike the specification of Lehnert's, not only is there a representation of the fabula, but there is a precise indication of what text segment a plot unit applies to. There is alas no automatic capability for annotation of SIGs.

4.1.6.5 Narrative Event Chains

The problem with representations based on goals, beliefs, and plans, as well as plot units, is that they are hard to infer, especially when the story is not about such plans. It may be more feasible, from a practical standpoint, to view the extraction of plots as a summarization problem, addressing plot as an event summary rather than a structure based on plot units that involves referencing the implicit goals and beliefs of the characters.

Here the notion of a Narrative Event Chain (NEC) of Chambers (2011) becomes relevant. An NEC is a temporal sequence (actually, a partial temporal ordering) of narrative events that share a common protagonist. NECs are inspired by, but less expressive and computationally expensive than scripts. However, they can be rather long, and can involve incidental events. More importantly, they can be extracted accurately from text, as shown in an evaluation involving completion of chains by Chambers (2011) as well as examples run with GPT-4o in this book.

Compressed Event Chains

Although NECs have not been used to address problems of plot, my idea is to compress these event chains, excising less salient events without exceeding the desired compression rate, so as to provide a plot summary. This also addresses my earlier concern about the lack of specifications of plot length. Such an approach can be formulated as a knapsack problem of picking a subset of the NECs such that their total length is less than a limit but where their total value, i.e., salience, is maximized. (This knapsack problem can in turn be solved by the mathematical optimization method known as dynamic programming.) To render the chains, we integrate them together to produce a summary. The summary can then be smoothed by referencing a particular temporal ordering template, as well as the use of coreference.

Turning to our *Engagement* story in Example 4.10, the NECs are as follows:

Example 4.13
NECs(J): M accepted J's engagement ring; Thrilled; Found out about M's father's biz; Felt torn; Arrested M's father.

NECs(M): Accepted J's engagement ring; Called off their engagement.

Summary: Mary accepted John's engagement ring; John arrested Mary's father; Mary called off their engagement.

Smoothed Summary: Mary accepted John's engagement ring. Then, when John arrested her father, she called off their engagement.

The approach of Lehnert (1981) to summarize plots involves deriving a graph of plot units which are connected together based on characters sharing common affects, and then selecting the most salient ones based on connectivity patterns in the graph, followed by text smoothing operations. The summary suggested by Lehnert (1981) is:

4.1 Introduction   143

Example 4.14
Smoothed Summary: Because John arrested Mary's father, she called off their engagement.

We will go into the details of NECs in Chap. 5, which in turn requires a richer NarrativeML DTD. As a preview, however, the reader should rest assured that LLMs can quite easily generate NEC-based summaries. For example, consider *The Wily Lion*, with the following prompt and response from GPT-4o:

Example 4.15

> **User Prompt**: Here is a DTD for NarrativeML 2.2 and an example annotated story.... Please generate the NarrativeML for this new story.... Then generate a summary based on fusing the NarrativeML NECs for each character, without using the source text.
> **Machine**: Driven by hunger, a Lion avoided attacking a horned Bull and chose deception instead. He flattered the Bull, who was persuaded to remove his horns. Left defenseless, the Bull was killed by the Lion.

I would have preferred to see "remove his horns" replaced with "have his horns cut off", but otherwise the summary in itself seems fine.

4.1.7 Comparison of Plot Models

We are now in a position to summarize our comparison of selected plot models, shown in Table 4.2. Here reliability has to do with annotation reliability. It can be seen that no single model has at present all of the desired characteristics.

Table 4.2 Comparisons of plot models

Model	Elements	Reliability	Computability
Narratemes	Events, character types	Low	Generation
Plot units	Affect, causal links	Good (affect)	Average (causal link)
DPF	Beliefs, preferences	Unknown	Unknown
SIG	Beliefs, goals, affect, time	Good	Unknown
NEC	Events, time	Unknown	Good

4.1.8 Narratological Implications

The creation of plots for narratives, whether based on computational approaches such as Plot Units, SIGs, and/or NECs, offers interesting explorations relevant to the interests of humanities narratology:

- Finding stories with similar plots. Consider again Aesop's *The Fox and the Crow*, translated by Jones (1912), shown in Example 4.16.

Example 4.16
A Crow was sitting on a branch of a tree with a piece of cheese in her beak when a Fox observed her and set his wits to work to discover some way of getting the cheese. Coming and standing under the tree he looked up and said, "What a noble bird I see above me! Her beauty is without equal, the hue of her plumage exquisite. If only her voice is as sweet as her looks are fair, she ought without doubt to be Queen of the Birds." The Crow was hugely flattered by this, and just to show the Fox that she could sing she gave a loud caw. Down came the cheese, of course, and the Fox, snatching it up, said, "You have a voice, madam, I see: what you want is wits."

As Elson (2012) shows, the results from automatic comparison of this fable with *The Wily Lion* at the level of their SIG annotations confirm our narratological intuitions about the reasons for their similarity: "the fox is like the lion and the crow is like the bull, in that in both stories, one is an inciting agent who devises a plan that has the other agent devising and executing its own plan. This particular analogy is further enriched by the aligned use of the flatter predicate in the plan" (ibid., p. 230).[3] Comparison of stories through NECs is even simpler, and is a clear possibility for future research. Altogether, such comparison algorithms allow for far more intelligent searching for similar stories than has been possible earlier. Such searches can greatly aid the narratologist in making claims to support theory development. They can also help clarify what it is we mean when we attribute a common plot to different narrative versions, say, of fairy tales like *Little Red Riding Hood*.

- Identifying causes for ambiguous contexts. Ambiguity may arise for a variety of reasons: (i) deliberate authorial ambiguity; (ii) genuine ambiguity due to lack of information, language, etc.; or (iii) a reflection of the inherent subjectivity of certain types of interpretation related to agents' motives, feelings, and beliefs. These missing values are often a cue for the narratologist to examine further the causes and significance of the ambiguity. The provision of numerous such contexts by a system can be a useful aid.

---

[3] My experiments show that LLMs are also able to carry out similar comparisons of the story pair, though at a lower level of abstraction, capturing details of the plot rather than the ownership of plans.

4.1 Introduction

- Dealing with goals related to non-existent characters, whom we have encountered in narratives like *A Beautiful Mind*, *The Continuity of Parks*, *The Thousand and One Nights*, etc. The plot model used, whether SIGs or NECs, must allow for agents whose `exists` flag is false, who are accessible only to certain agents. Metalepsis, for example, has not, to my knowledge, been addressed in the context of any of the models we have seen here.
- The above plot models have not been extended to handle narrator goals. Without representing such goals, the ending in many stories makes no sense. For example, in *Tess of the d'Urbervilles*, Tess, towards whom the narrator is increasingly sympathetic, is executed at the end, an outcome that would leave the reader sad. Why end on such a tragic note? As Howe (1967) has argued, it is clearly to underline how the innocent are made victims of morally hypocritical social institutions in Victorian society.
- Plots need to be tied to reader evaluations, both for narrative understanding and generation. Consider this micro-narrative discussed in the narratological literature by Kukkonen (2014), based in turn on Brewer and Lichtenstein (1981) and Sternberg (1993):

1. Butler puts poison in wine.
2. Butler carries wine to Lord Higginbotham.
3. Lord Higginbotham drinks wine.
4. Lord Higginbotham dies.

As Kukkonen points out, if readers are given only sentence 4 as the entire narrative, they would be surprised, as there is no indication of the cause of the unfortunate event. If given sentences 2–4 as the entire narrative, it would stoke their curiosity, as they are trying to find an earlier cause (and, one would think, are primed to expect poison). Whereas if they are given sentences 1–3 only, they would experience suspense, which involves their guessing what might happen next. She goes on to argue that these and other examples indicate that as a narrative progresses, readers' prior probabilities are revised, and usually improved, based on the information from the plot, in turn suggesting a Bayesian model of reader evaluations tied to plot.. For example, "More information about the butler, his whereabouts during the Lord's death and his ordering arsenic the day before, will make the possibility that he kills the Lord more and more likely." (Kukkonen 2014, p. 726). The relationship between plot, reader evaluations, and surprise, curiosity, and suspense, especially conceived of within a Bayesian framework, is extremely interesting and deserves to be explored in computational terms.

One point worth making here is the use of commonsense resources to infer causal relations for plot discovery. GLUCOSE (Glucose 2020) is a commonsense resource consisting of the ROC stories annotated via crowdsourcing with ten dimensions of causal information.

For example, the sentence "Gage turned his bike sharply" will have a causal or enablement link to the antecedent event from the preceding sentence "A car turned in front of him", as well as the drive that motivated the event (Gage wants safety), the enabling location state

(Gage was close to the car), the enabling possession state (Gage possesses a bike), a resulting change of location (Gage is further away from the car). These annotations are semi-structured and expressed in logical relations that are mapped to syntactic relations.[4] Training a model on the GLUCOSE knowledge base and then giving it a sentence in a story like the above one mentioning 'Gage', results in the model predicting such causal relationships. The model has been successfully evaluated by humans in terms of its ability to answer questions about causal relations in a story. In effect, the GLUCOSE knowledge is being used to infer the structure of the plot, albeit at a local level.

4.2 NarrativeML

We now add a PLOT that is bare-bones, integrating only character GOALs and NECs. It will be elaborated based on specific causal links that we will discuss in the context of narrative generation in Chap. 6.

Example 4.17

```
<?xml version="1.0" encoding="UTF-8"?>
<!DOCTYPE NarrativeML [
<!ELEMENT NARRATIVE (#PCDATA | NARRATOR | AUDIENCE | CHARACTER |
 EVALUATION | EVENT | TIME | SEGMENT | PLOT | GOAL | NEC)*>
...
<!ELEMENT PLOT EMPTY>
<!ATTLIST PLOT id ID #REQUIRED>
<!ATTLIST PLOT NECS IDREFS #IMPLIED>
<!ATTLIST PLOT GOALS IDREFS #IMPLIED>
...
<!ELEMENT GOAL EMPTY>
<!ATTLIST GOAL id ID #REQUIRED>
<!ATTLIST GOAL parent IDREF #IMPLIED>
<!ATTLIST GOAL character IDREF #IMPLIED>
<!ATTLIST GOAL leaf (true | false) #REQUIRED>
<!ATTLIST GOAL events IDREFS #IMPLIED>
...
<!ATTLIST EVENT goals IDREFS #IMPLIED>
...
<!ELEMENT NEC EMPTY>
<!ATTLIST NEC id ID #REQUIRED>
```

---

[4] Thus the first rule above is really that an entity x (a car) turning in front of a vehicle y (a bicycle) belonging to z (the biker) causes or enables z (the biker) to turn y (the bike) away from x (the car), with those slots mapped to Prop Bank style predicates and arguments corresponding to syntactic relations in the sentence.

```
<!ATTLIST NEC entity IDREF #IMPLIED>
<!ATTLIST NEC events IDREFS #IMPLIED>
]>
```

4.3 Conclusion

Plot turns out to be the central concept of narrative. The classical narratological notions of plot arise from widely accepted causal definitions. The picture we have painted of plot is that there is no one representation; they have different levels of granularity and come with different tradeoffs. Nevertheless, these representations have been investigated in considerable depth. Propp's narrative functions, which have been applied to folk tales from several cultures, have attracted computational interest, with interesting results. More explicitly computational models, all of which are graph-based, including Plot Units, Doxastic Preferences, Story Intention Graphs, and Narrative Event Chains. LLMs can be used to good effect in computing some of these. Comparisons of different models suggest some of the computational limitations and the ramifications for humanities narratology. A light plot-related layer was added to NarrativeML, involving character goals and NECs, that will be elaborated further in subsequent chapters. When we discuss Story Generation in Chap. 6, we will see even more flexibility in the representation of plot.

References

Abrams, Meyer H. (1993). *Glossary of Literary Terms, 6th edition*. Fort Worth: Harcourt Brace.
Andel, Kevin M. van (2010). *Formalizing TV Crime Series: Application and Evaluation of the Doxastic Preference Framework*. Bachelor's thesis, Faculty of Science, University of Amsterdam.
Aristotle (1932). *Poetics*. Project Gutenberg, Number 1974.
Bamman, David and Noah A. Smith (2013). "New Alignment Methods for Discriminative Book Summarization". URL: https://www.cs.cmu.edu/~dbamman/booksummaries.html.
Bartlett, Frederic (1932). *Remembering*. Cambridge, UK: Cambridge University Press.
Bod, Rens et al. (2012). "Objectivity and reproducibility of formal narrative representations or annotations: Propp's functions and narrative summarization". In: *Proceedings of The Third Workshop on Computational Models of Narrative*. Cambridge, MA, pp. 17–21. URL: https://narrative.csail.mit.edu/cmn12/proceedings.pdf.
Brewer, William and Edward Lichtenstein (1981). "Event Schemas, Story Schemas and Story Grammars". In: *Attention and Performance. Volume 9*. Ed. by John Long and Alan Baddeley. Hillsdale, NJ: Lawrence Erlbaum, pp. 636–679.
Campbell, Joseph (1949). *The Hero with a Thousand Faces*. New York: Harper and Row.
Chambers, Nathanael (2011). "Inducing Event Schemas and their Participants from Unlabeled Text". PhD thesis. Department of Computer Science, Stanford University.
Cullingford, Richard E. (1978). *Script application: computer understanding of newspaper stories*. Tech. rep. Research Report 116. Computer Science Department, Yale University.

DeJong, Gerald F. (1982). "An overview of the FRUMP system." In: *Strategies for Natural Language Processing*. Ed. by Wendy G. Lehnert and Martin H. Ringle. Hillsdale, NJ: Lawrence Erlbaum, pp. 149–176.

Dundes, Alan (2005). "Folkloristics in the Twenty-First Century (AFS invited presidential plenary address)". In: *Journal of American Folklore* 118.470, pp. 385–408. URL: http://dx.doi.org/10.1353/jaf.2005.0044.

Dundes, Alan (1963). *The Morphology of North American Indian Folktales*. Helsinki: Suomalainen Tiedeakatemia.

Elson, David K. (2012). "Modeling Narrative Discourse". PhD thesis. Department of Computer Science, Columbia University.

Finlayson, Mark (2016). "Inferring Propp's Functions from Semantically Annotated Text". In: *Journal of American Folklore* 129.511, pp. 55–77. URL: https://users.cis.fiu.edu/~markaf/doc/j4.finlayson.2016.jaf.129.55%5C_archival.pdf.

Forster, E. M. (1956). *Aspects of the Novel*. New York: Harcourt.

Freytag, Gustav (1900). *Freytag's Technique of the Drama: An Exposition of Dramatic Composition and Art*. Chicago: Scott, Foresman.

Gervas, Pablo and Gonzalo Mendez (2024). "Tagging Narrative with Propp's Character Functions Using Large Language Models". In: *Proceedings of the Text2Story'24 Workshop, Glasgow (UK)*. URL: https://ceur-ws.org/Vol-3671/paper12.pdf.

Goyal, Amit, Ellen Riloff, and Hal Daume (2010). "Automatically producing plot unit representations for narrative text". In: *Proceedings of the Conference on Empirical Methods in Natural Language Processing (EMNLP'2010)*. Cambridge, MA, pp. 77–86. URL: https://aclanthology.org/D10-1008/.

Greimas, Algirdas Julien (1984). *Structural semantics: an attempt at a method*. Lincoln, Nebraska: University of Nebraska Press.

Horace (1976). *The Art of Poetry: An Epistle to the Pisos*. Project Gutenberg, Number 9175.

Howe, Irving (1967). *Thomas Hardy*. New York: MacMillan.

Jones, V. S. Vernon (1912). *Aesop's Fables: A New Translation*. New York: Avenel Books.

Kukkonen, Karin (2014). "Bayesian Narrative: Probability, Plot and the Shape of the Fictional World". In: *Anglia* 132.4, pp. 720–739. URL: https://www.duo.uio.no/bitstream/handle/10852/54629/1/01%5C%2BKukkonen%5C%2BBayesian%5C%2BNarrative.pdf.

Laurel, Brenda (1993). *Computers as Theatre*. Boston, MA: Addison-Wesley Longman Publishing.

Lehnert, Wendy G. (1981). "Plot Units: a narrative summarization strategy". In: *Strategies for Natural Language Processing*. Ed. by Wendy G. Lehnert and Martin H. Ringle. Hillsdale, NJ: Lawrence Erlbaum.

Lowe, Benedikt (2010). "Comparing formal frameworks of narrative structures". In: *Computational Models of Narrative: Papers from the 2010 Association for Advancement of Artificial Intelligence (AAAI) Fall Symposium*. Washington, DC: AAAI Press. URL: https://eprints.illc.uva.nl/id/eprint/407/1/PP-2011-03.text.pdf.

Mateas, Michael (2000). "A neo-Aristotelian theory of interactive drama." In: *Working Notes of the American Association for Artificial Intelligence (AAAI) Spring Symposium on Artificial Intelligence and Interactive Entertainment*. Washington, DC: AAAI Press. URL: https://aaai.org/papers/0011-ss00-02-011-a-neo-aristotelian-theory-of-interactive-drama/.

Mateas, Michael and Andrew Stern (2005). "Structuring content in the Facade interactive drama architecture." In: *Proceedings of the First Conference on Artificial Intelligence and Interactive Digital Entertainment (AIIDE'2005)*. Washington, DC: AAAI Press, pp. 93–98. URL: https://ojs.aaai.org/index.php/AIIDE/article/view/18722.

Mostafazadeh, Nasrin, Aditya Kalyanpur, et al. (2020). "GLUCOSE: GeneraLized and COntextualized Story Explanations". In: *Proceedings of the 2020 Conference on Empirical Methods in Natural Language Processing (EMNLP'2020)*. URL: https://aclanthology.org/2020.emnlp-main.370/.

References

Murray, Janet H. (1997). *Hamlet on the Holodeck: The Future of Narrative in Cyberspace*. New York: The Free Press.

Propp, Vladimir (1968). *Morphology of the Folktale*. Austin: University of Texas Press.

Ramanujan, A. K. (1991). *Folktales from India, Oral Tales from Twenty Indian Languages*. New York: Pantheon.

Ramanujan, A. K. (1997). *A Flowering Tree and Other Oral Tales from India*. New Delhi: Penguin.

Ryan, Marie-Laure (1991). *Possible Worlds, Artificial Intelligence and Narrative Theory*. Bloomington: Indiana University Press.

Schank, Roger C. and Robert P. Abelson (1977). *Scripts, Plans, Goals and Understanding: An Inquiry into Human Knowledge Structures*. Hillsdale, NJ: Lawrence Erlbaum.

Scheidel, Antonia and Thierry Declerck (2010). "APftML – Augmented Proppian fairy tale Markup Language". In: *First International AMICUS Workshop on Automated Motif Discovery in Cultural Heritage and Scientific Communication Texts, Hungary*. Szeged University, Szeged. URL: https://www.dfki.de/fileadmin/user_upload/import/5050_AMICUS-APftML.pdf.

Sternberg, Meir (1993). *Expositional Modes and Temporal Ordering in Fiction*. Bloomington: Indiana University Press.

Yarlott, W. V. H. and Mark Finlayson (2016). "ProppML: A Complete Annotation Scheme for Proppian Morphologies". In: *7th International Workshop on Computational Models of Narrative (CMN'16)*. URL: https://core.ac.uk/download/pdf/62923031.pdf.

Time 5

> *"I wonder", said Ada. "I wonder if the attempt to discover those things is worth the stained glass. We can know the time, we can know a time. We can never know Time. Our senses are simply not meant to perceive it. It is like —".*
>
> —Vladimir Nabokov, Ada

5.1 Introduction

Time is central to narrative, but it is a somewhat mysterious concept. In real life, time is not directly perceived, though we can measure its passage in various ways. We experience time subjectively, with the positions of events in time changing dynamically relative to the speaker. The subjective notion of time gives rise to the distinctions of tense in human languages. With the past tense, the event occurs prior to the time of the narrating event, called the speech time; with the present, it occurs roughly at the speech time; and with the future tense, the event time is later than the speech time. Note that when we speak of time, we are often talking about events in time, so temporal positioning of those times and events becomes crucial. It turns out that even very simple stories can be very rich in temporal information. Sometimes the temporal relations are stated explicitly, but more often not; and at times it is hard to tell from the narrative in what order events actually occur.

In what follows, we will first explore the clues to time provided through linguistic information, particularly the systems of tense and aspect that help position events in time. We then examine the orders in which events can be narrated, which may be different from the order in the fabula; for example, a narrative may start with the most recent event first. After discussing the seven types of ordering posited by Genette (1980), we turn to psychological experiments on human processing of temporal ordering in narrative. We then explain how to annotate times and temporal relations in stories, relying on an underlying AI reasoning formalism called the interval calculus. The annotation can also capture situations where times and events are remembered or imagined, leading to the use of subordination relations. While human annotators agree well on how to annotate the times using the formalism, they may disagree on what temporal relations hold between certain pairs of events. We then point to research efforts aimed at remedying this problem. Next, we examine how machines can automatically understand the temporal structures in narratives. We see to what extent LLMs can infer the correct underlying temporal relations in stories exhibiting the seven different orderings. We also discuss historical work on inferring the temporal structure of entire stories. We then turn to two topics which haven't been given much attention in computational approaches: inferring tempo, and reasoning about durations of events. We then extend NarrativeML to take into account these extensive temporal facets of narrative, before concluding.

5.2 Temporal Positioning: Tense and Aspect

Tense is a grammaticalized expression of temporal location, and is indicated by the verb and auxiliaries in English, but also by other parts of speech, including, for a wide variety of languages, noun phrases (Comrie 1986, NordlingerSadler 2004). Languages also vary in terms of the number of tenses. (For example, the Bantu language ChiBemba has four past tenses and four future tenses.) Some languages lack such grammaticalized expressions altogether, allowing time and date locutions and aspectual markers to play a greater role along with context, as in the case of Mandarin Chinese, as discussed by Lin (2003). Finally, there are languages like Burmese, which fails to distinguish past, present, and future in the absence of time and date locutions, instead using *realis* and *irrealis* particles to differentiate between ongoing or past events, and others.

Other linguistic mechanisms to express time include grammatical aspect, which represents internal phases of an event, indicating whether it is terminated or completed at a particular time (the perfective aspect, as in "We hunted yesterday"), or whether it is ongoing at the time (the imperfective aspect, e.g.,"We were hunting yesterday"). Temporal locutions expressed by adverbials, noun phrases, and prepositional phrases in English convey dates and times as well as temporal relations.

Unlike the purely mathematical abstraction of time as an infinite succession of infinitesimal instants, natural languages conceptualize time in terms of successions of events that each have a certain duration, irrespective of whether that duration is made explicit. Therefore events need to be treated as occupying finite intervals of time; a character can blink, or have lunch at noon, but in either case the author and reader are construing the event as taking time.

The linguistic devices of tense and aspect allow narrators and characters to express, from their viewpoint, the position, duration, and tempo of events in time. Narrative generation systems that vary the tense or aspect thus have to be cognizant of the way such devices operate in a given language. Narrative understanding systems also need to be aware of them in mapping from text to underlying representations such as fabulae. From the standpoint of computational narratology, it is crucial to formalize the representation of time. For historical background on time in language and computation, see Mani et al. (2005).

5.2 Temporal Positioning: Tense and Aspect

Let us consider a few linguistic examples in which language communicates position information through tense and aspect about the order in which events occur in a narrative.

In Example 5.1, the narrative convention of simple past tense events being described in the order in which they occur is followed.

Example 5.1
Mary stood up. John greeted her.

However, it is overridden in Example 5.2.

Example 5.2
Yesterday, John fell and broke his leg. Mary pushed him.

In Example 5.3, the perfective form indicates that the drinking was completed.

Example 5.3
Mary entered the room. She had consumed two bottles of wine.

Whether the event is being presented as a state, i.e., stative or not, also matters. Example 5.4 has the state of being seated overlapping temporally with the event of entering.

Example 5.4
John entered the room. Mary was seated behind the desk.

Let's zoom into the details of the times in a slightly more extensive narrative. Consider the following ROC story (Mostafazadeh et al. 2016):

Example 5.5
I had always been afraid of dogs. So I felt nervous while walking home one night. I saw what looked like a dog on someone's porch. And I started to run. The next day, I realized it was a statue.

None of the times in Example 5.5 are anchored to a calendar. The past perfect tense is used to indicate that until some time $t1$ earlier than the speech time $t2$ of the narrator speaking, the narrator had been afraid of dogs. We treat 'afraid of' as a state $e1$, which is a variety of event. At a night time $t3$ between $t1$ and $t2$, s/he was walking ($e2$) home and feeling nervous ($e3$). At $t4$, one day after $t3$, but still before $t2$, the realization (state $e4$) took place.

It can be seen that such a simple story, which even a child could understand, is very rich in temporal information. Now consider the events. When do they occur? Clearly, event $e1$ ('afraid of', in past perfect tense) is either during or co-extensive with the time interval ending in $t1$ (we will shortly talk about intervals). Events $e2$ ('walking', in past progressive tense) and $e3$ ('felt nervous', in past tense) are during $t3$, and $e2$ is either during or co-extensive with $e3$. As for event $e4$ ('realized', past tense), it is during $t4$.

The above analysis, which is more formalized than what a child would say about the relationships, is what I have called a 'conservative interpretation' in my book on time in narrative (Mani 2010), and is related to narratological ideas like Ryan's Principle of Minimal Departure (Ryan 1991). Of course, this commonsense analysis glosses over further fine-grained details that a linguist might want to analyze further. The use of the progressive tense form 'was walking home' means that the narrator did walk, even if there were gaps when she stopped to check her phone or whatever, and it does not commit to her having reached home.

In Example 5.5 we came across the expression 'had always been'. Like 'used to', these are habitual expressions, which in Romance languages are expressed using the imperfect tense. One common narrative device is to express *habituality* in terms of a set of times, and then to zoom (i.e., drill-down) in to a particular scene as if it exemplified those times, though clearly the particulars of the scene are far too specific to be repeated. Here are the opening lines of Kafka's *The Hunger Artist*:

Example 5.6
During these last decades the interest in professional fasting has markedly diminished. It used to pay very well to stage such great performances under one's own management, but today that is quite impossible. We live in a different world now. At one time the whole town took a lively interest in the hunger artist; from day to day of his fast the excitement mounted; everybody wanted to see once a day; there were people who bought season tickets for the last few days and sat from morning till night in front of his small barred cage; even in the night time there were visiting hours, when the whole effect was heightened by torch flares; on fine days the cage was set out in the open air ...

It is worth noting that habituality and quantification can be used to express deliberate vagueness about when things happen. In his novel *Austerlitz*, W. G. Sebald uses iterated events and ranges of time to create an impression of repeated journeys in search of something; the times and durations of these journeys are left deliberately vague.

Example 5.7
In the second half of the 1960's I traveled repeatedly from England to Belgium, partly for study purposes, partly for other reasons which were never entirely clear to me, staying sometimes for just one or two days, sometimes for several weeks.

5.3 Narratological Distinctions

5.3.1 Narrative Time

The time of the narrating event is known in linguistics as the **speech time**, which is distinguished from the **event time** of the events being talked about in the narrative. Thus,

5.3 Narratological Distinctions

when someone says, "I skipped work yesterday," the event time of skipping work is one day before the speech time. A third category, of **reference time**, following (Reichenbach 1947), is exemplified by expressions like "Tuesday," "the next week," etc., but can also be implicit in the text. In the sentence "The suspect had already confessed," the reference time is unspecified, and the event of confessing precedes the reference time, with both these times preceding the speech time. In Reichenbach's scheme, this ordering captures the semantics of the past perfect tense. In the simple past tense, the reference and event times coincide with and precede the speech time.

It is traditional in narratology to speak of the narrative being before, at, or after the time of narration. This gives rise to three different narrative time relations, respectively: **subsequent**, **simultaneous**, and **prior**, as Genette (1980) also points out.[1] These relations determine in turn the basic tense (past, present, and future) used by the narrator.

Returning to the distinction between fabula and discourse, an important temporal distinction is between **story time** "the duration of the purported events of the narrative" and **discourse time**, "the time it takes to peruse the discourse" (Chatman 1980, p. 62). In computational terms, the story time of a narrative can be defined more precisely as the length of the time interval between the earliest and latest event in the narrative fabula. This may be based on predications made in the narrative as to how long particular events may have lasted, or else it may have to be derived from commonsense intuitions. The discourse time, on the other hand, can be measured in terms of the length of the recounting of the discourse of the narrative. This length can be in words in the case of text, and feet or frames of film in the case of a movie. As we will see in Sect. 5.7.2, the ratio between the two influences the pace of the narrative.

5.3.2 Narrative Order

Consider the relation between the **order** of narration in the discourse (i.e., text order), in relation to actual order of events in the fabula. Genette (1980) identifies seven types of ordering:

Achrony
Narrating events such that their order is left unclear, as in Robbe-Grillet's *Jealousy*.

Example 5.8
Besides, she was no longer facing Franck at that moment. She had just moved her head back and was looking straight ahead of her down the table, toward the bare wall where a blackish spot marks the place where a centipede was squashed last week, at the beginning of the month, perhaps the month before, or later.

---

[1] Genette further distinguishes a fourth time relation of *interpolated* narration, as exemplified by epistolatory novels, but this involves a combination of simultaneous and prior narration.

Analepsis
Flashback, i.e., narrating an event that occurred earlier than the current point in the narrative.

This will be intimately familiar to the reader, and we already saw an example in Benjy's flashback in the Faulkner extract in Chap. 3, Example 3.15.

Chronicle
Narrating events in their order of occurrence. This is the most common and easiest to follow.

Prolepsis
Flashforward, i.e., narrating an event in advance of the current point in the narrative. This a staple of TV serials as well, aimed at retaining audiences, creating suspense and foreshadowing future plot developments. Example 5.9 illustrates this:

Example 5.9
At that very first high-school date she had a sudden image of the two of them sitting together in their old age in rocking chairs in the front porch of her grandparents' home. She knew then that he was the right choice.

Retrograde
Narrating events in reverse order of their occurrence, as attempted in the novel *Time's Arrow* by Martin Amis. It is relatively rare to find this on a macro-level across large swathes of text, as it becomes hard for a human to process. But it is fairly common on a local level.[2] Here is a constructed example adapted from our ROC spaghetti story:

Example 5.10
Tina was touched that her boyfriend had pretended that the spaghetti she had cooked for him was good, to spare her feelings. She had tried it herself and realized it was disgusting after he had said it was good, upon eating the whole plate.

Syllepsis
Narrating events based on an ordering that is non-temporal, e.g., spatial. In literature, it is exemplified by George Perec's novel *Life: A User's Manual*. The pattern is closely related to achrony, but differs from it in having a spatial arrangement. Here is another constructed example:

Example 5.11
The gunshot was drowned by the sound of the midnight train pulling away. In the hotel room, she hesitated as she took the revolver out of her suitcase. In the restaurant, he promised her they would be together forever once they reached Venice.

---

[2] In traditional rhetoric (Quinn 1982), this local temporal reversal is called *hysteron-proteron*, e.g., "Let us die and rush into the heart of battle".

Zigzag
Where the narrator alternates between two times, e.g., 'now' and 'once', as in Example 3.18 from Proust's *Jean Santeuil*, repeated here:

Example 5.12
Sometimes passing in front of the hotel he remembered the rainy days when he used to bring his nursemaid that far, on a pilgrimage. But he remembered them without the melancholy that he then thought he would surely some day savor on feeling that he no longer loved her.

Here we have the events and times in Jean's memories (of a past involving rainy days with his nursemaid) *subordinated* to the act of remembering, and his thoughts (at that past time) of the future, with the events and times of that future subordinated to the act of thinking. As Genette (1980) points out, this pattern where the "future has become present but does not resemble the idea of it that one had in the past" is a form that appears often in the Proust's *magnum opus*.

5.4 Temporal Narrative: Human Processing Results

Experiments by Zwaan (1996) have shown that in reading narrative passages, readers expect that successive sentences will describe chronologically successive and contiguous (i.e., temporally adjacent) events. Deviations from this will result in delays in processing information. For example, the study discovered that sentences with a time shift between events (e.g., 'an hour later') take longer to read, and result in subjects taking a longer time to answer questions about whether a particular word occurred in the story, compared to when there isn't such a shift. Events separated by a time shift were less strongly connected in long-term memory than those that were not separated by a narrative time shift. Also, as shown in those experiments, when processing a narrative sequence of immediately successive events without a time shift, readers took longer to access events that, although mentioned recently, were temporally somewhat remote from the current narrative 'now'. This temporal distance effect was absent when the text had a time shift.

Studies by Graesser et al. (2002) of reading comprehension suggest that humans also track causal relations between events, which involve event ordering in the fabula as well as characters' goals. Taken together, the body of psychological experiments supports the hypothesis that readers build cognitive models in their minds of the situation described by the narrative, including representations of whether events are before or after each other, and how far apart they are in time. In other words, readers construct timelines, with a strong preference for simple ones.

A neuroscience study by Grall et al. (2023) had subjects watch the movie *500 Days of Summer*, a romcom which follows a couple through 500 days of their relationship. The movie has a highly, non-linear narrative, with the narrative order not being chronological and with

Table 5.1 Interval calculus relations for annotation

BEFORE	xxxx　　yyyy
MEETS	xxxxyyyy
SIMULTANEOUS	xxxx yyyy
OVERLAPS	xxxx 　yyyy
BEGINS	xxxx yyyyyy
DURING	xxxx yyyyyyyy
ENDS	xxxx yyyyyy

many instances of analepsis and prolepsis. To keep the movie viewers from being totally confused, the somewhat heavy-handed tactic of time cards with the upcoming day number occur at these jumps. The study's results reveal that certain brain regions engage in on-the-fly temporal unscrambling to represent information chronologically. Days that are closer together chronologically are represented more similarly no matter when they are presented in the movie. Overall, the study emphasizes the importance of chronological information when the brain is encoding narrative events.

5.5 Temporal Representation

5.5.1 Interval Calculus

5.5.1.1 Introduction

In the temporal representation used here, events are treated as time intervals (as are times), and time intervals are related by relations from the interval calculus (Allen 1983, 1984), shown in Table 5.1.[3]

In addition to the seven relations shown in Table 5.1, there are six other inverse relations, together combining to provide a calculus of 13 relations:

- AFTER (inverse of BEFORE);
- MET BY (inverse of MEETS);
- OVERLAPPED BY (inverse of OVERLAPS);
- BEGUN BY (inverse of BEGINS);

---

[3] It is of course possible to further distinguish events from intervals, as done by Allen (1984), so that events can occur during an interval, either for the entire interval or for some part of it.

- CONTAINS (inverse of DURING); and
- ENDED BY (inverse of ENDS).

Annotating a narrative with the relations in Table 5.1 is not the end of the story. Some of the temporal relations may be absent in the text; the most extreme case of this being achrony. When human readers carry out the annotation, they may differ in what they include or leave out, especially when the text is longer or there are plenty of relations to annotate. If event A is before event B which in turn is before event C, an annotator may record that, while omitting to record that A is also before C. Another annotator may record that missing fact, resulting in discrepancies across annotators. AI reasoning can help bridge these differences, but temporal reasoning comes with its own challenges. The next section discusses these issues in technical detail, beginning with a story by Chekhov. It may be skipped by readers who aren't interested in the technical specifics, which also involve some mathematics.

5.5.1.2 Interval Calculus Reasoning Challenges*

In narratives longer than our constructed micro-stories, only some of the events will be related explicitly by a constraint represented as one of the interval calculus's 13 temporal relations. Others may not be related by a single one of the 13, i.e., they will be expressible as the disjunction from among these 13 relations, e.g., the constraint 'A BEFORE B or A MEETS B'. This is not only the case for instances of achrony, where the ordering of most of the events is deliberately omitted; most texts will consist of such partial orderings. Note that for a *base* set of 13 temporal relations, there is a larger *underlying* set of 2^{13} possible disjunctions of relations, any one of which can hold between a pair of events.

As an example, consider Chekhov's story *The Lady with the Pet Dog*. The event X of Gurov seducing Anna is before the event Y of his confronting her at the theater. This can be expressed as a metric (quantitative) constraint: $(Y_{start} - X_{end}) \geq 0$. Now, Gurov sees Anna off at the station in Yalta at the beginning of autumn, several weeks after first meeting her, and then visits her in the theater that very December. Thus $(Y_{end} - X_{end}) \leq 5$ months. A general procedure for representing numeric constraints like these in the interval calculus is found in the *Allen-to-metric* method described by Kautz and Ladkin (1991).

Let us zoom in on the interval calculus itself. All the 13 relations can in fact be expressed in terms of the relation MEETS. For example, A is BEFORE B if there exists an interval C such that A MEETS C and C MEETS B. To learn more about the reduction of all temporal relations to MEETS, a paper by Allen and Hayes (1985) formalizes the interval calculus in first-order logic based on five axioms which specify that the meetings of intervals are unique, that pairs of intervals meet in linear order, that the union of every pair of intervals exists, and that time is infinite while intervals are finite. The relations in the interval calculus can in turn be composed together using a composition operator ∘, for example, so that if A is BEFORE B and C DURING B, A is BEFORE C. In other words, BEFORE(A, B) ∘ DURING(C, B) = BEFORE(A, C). Here, the composition table that defines ∘ has $13^2 = 169$ entries.

A temporal representation for a narrative can thus be viewed as a directed graph $(V, E, \circ)$, which we call here a **temporal graph**, where V is a set of nodes representing events and times, E is a set of edges each of which represents a constraint C_{ij} between a pair of nodes i and j, and $\circ$ is the composition function.

A key computational problem in dealing with temporal graphs is determining if the graph is free of inconsistencies, such as A BEFORE B and B BEFORE C along with the clashing constraint C BEFORE B. These inconsistencies can easily arise when a human or machine annotates a narrative, as local decisions often come into conflict with global constraints. There are also cases where the narrative itself is temporally inconsistent, as when there is a cycle in time. This is not that rare in mythology, as discussed in Mani (2010).

To flag inconsistencies, a system can verify whether every pair of nodes in the temporal graph that are known to be consistent can be extended to a triple of nodes that are consistent. The key step in the algorithm is to compute, for all triples of nodes i, j, k in the graph, the value of temporal constraint C_{ij} to be the old value of C_{ij} intersected with the composition $C_{ik} \circ C_{kj}$. This *path consistency* algorithm thus computes the transitive closure of the graph. However, path consistency is a necessary condition but not in general sufficient for establishing consistency; Vilain et al. (1989) found pathological inconsistencies involving quadruples of nodes that a path-consistency algorithm (such as the one in Allen (1983)) will not detect. One can get around such pathologies, which one may or may not run into in practice, by restricting the interval calculus severely. A restriction which uses all the 13 base relations, but only a proper subset of 10% of the underlying relations, was discovered by Nebel and Burckert (1995).

5.5.2 Temporal Annotation Scheme

We discuss here the TIMEX2 and TimeML annotation schemes, both of which have been imported into NarrativeML.

The TIMEX2 annotation scheme of Ferro et al. (2005) has been used to mark up time (and date) expressions in natural language with tags that indicate the extent and the resolved value of the time expression. A distinction is made between durative expressions, involving particular lengths of times predicated of events, and times that are treated in language as if they are points. For example, consider Example 5.13, which provides the inline XML.

Example 5.13

```
He finished a <TIMEX2 value="PT3H" anchorTimeID="t2">three-hour</TIMEX2>
meeting with the president <TIMEX2 value="1999-07-15">today</TIMEX2>.
```

Here 'three-hour' is marked as a period of time (PT) while 'today" is given a value, based on the reference time given by the document date. For times with fuzzy boundaries, such as 'winter', TIMEX2 introduces primitives (like WI). The time zone will of course vary based on geographic longitude and geopolitical region, and TIMEX2 also factors in different calendars. TIMEX2 annotation has been carried out on corpora in Arabic, Chinese, English, French, Hindi, Italian, Korean, Persian, Portuguese, Spanish, Swedish, Thai, and other languages, with detailed annotation guidelines available for Arabic, Chinese, English, and Persian. The scheme has been applied to news, scheduling dialogs, email, blogs, and other genres.

Annotators agree quite often on TIMEX2 tags: inter-annotator agreement is 85% F-measure on extent (where the tags start and end in the text) and 80% F-measure on time values for English in the 2004 TERN (Time Expression Recognition and Normalization) competition organized by the Automatic Content Extraction (ACE) program. That accuracy is acceptable for using the annotations as ground truth to evaluate systems. Some of the problem cases giving rise to disagreement include durative expressions where insufficient context is available. For example, in Example 5.14, from Ferro et al. (2005), the actual time of the shearing achievement is not specified, and as a result, TERN annotators differ in terms of when the 10 months of training ended.

Example 5.14
12-02-2000: 16 years of experience and 10 months of training are paying off for a sheep-shearer from New Zealand. Rodney Sutton broke a seven-year-old world record by shearing 839 lambs in nine hours.

TimeML, as described by Pustejovsky et al. (2005), is an annotation scheme for markup of events, times, and their temporal relations. It uses a TLINK tag to link events or times to other events or times, with the times marked up with an extension of TIMEX2 called TIMEX3.[4] Thus, Example 5.2 would be annotated as in Example 5.15. Here I am using logic instead of XML tags for ease of reading. It is extremely straightforward. BEFORE(e1, e2) means that e1 is BEFORE e2. DURING(e1, t1) means the e1 is DURING t1.

Example 5.15
Yesterday$_{t1}$, John fell$_{e1}$ and broke$_{e2}$ his leg. Mary pushed$_{e3}$ him.
BEFORE(e1, e2) & BEFORE(e3, e1) & DURING(e1, t1)
DURING(e2, t1) & DURING(e3, t1)

TimeML has been applied to news, medical narratives, accident reports, fiction, etc., and a version of it has been standardized as ISO-TimeML. Corpora annotated with TimeML (called *timebanks*) have been created for a variety of languages including English, by Pustejovsky et al. (2003), Catalan, by Sauri and Badia (2012), French, by Andre et al. (2011), German, by Spreyer and Frank (2008), Italian, by Caselli et al. (2011), Portuguese, by Costa and

---

[4] It is worth noting that TLINKs can be given a formal semantics, as by Bunt and Overbeeke (2008).

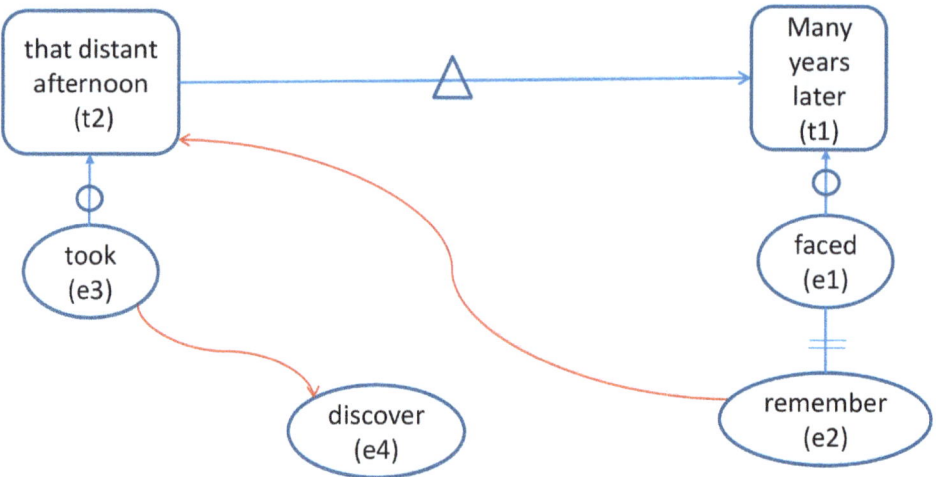

Fig. 5.1 Temporal graph for 1st sentence of *One Hundred Years of Solitude*

Branco (2012), Romanian, by Forascu et al. (2007), and Spanish, by Nieto et al. (2011), and other languages as well.

To see the associated temporal graph, let us examine more closely the famous opening sentence from *One Hundred Years of Solitude*, shown in Example 5.16.

Example 5.16
Many years later, as he faced the firing squad, Colonel Aureliano Buendia was to remember that distant afternoon when his father took him to discover ice.

A human-annotated temporal graph for the opening sentence is shown in Fig. 5.1.[5] Note that in TimeML, both events and times are treated as time intervals, and are given subscripts indicative of narration order, whereas the order of occurrence (the chronology) of the events is displayed left-to-right. In Fig. 5.1, the straight arrows with triangles indicate BEFORE relations (the interval at the arrow's tail being BEFORE the one at its head). The straight arrows with ovals indicate DURING relations, with the interval below being DURING the one above. The straight lines with a pair of horizontal bars indicate SIMULTANEOUS relations. As "that distant afternoon" is earlier in time than the "many years later," the computer can infer from the human annotation that Aureliano's father's taking him to discover ice (event $e3$) is BEFORE his facing of the firing squad (event $e1$).

Is such a temporal graph a suitable representation for a narrative's temporal properties? Certainly, it represents the relation between the ordering in the fabula (indicated by TLINKs)

---

[5] For higher resolution, the screen dump from the annotation tool has been redrawn here by hand.

5.5 Temporal Representation

and the ordering in the discourse (indicated by the ordering of event subscripts). All the seven orderings discussed by Genette can be expressed in this representation:

Achrony
Events with lower subscripts are not systematically related by temporal links to those with higher subscripts.

Analepsis
A chain of events with subscripts e_i, e_{i+1}, and (after a possible gap) e_{i+n}, such that e_{i+1} is BEFORE e_i which is BEFORE e_{i+n}.

Chronicle
All events with lower subscripts are related by BEFORE links to those with higher subscripts.

Prolepsis
A chain of events with subscripts e_i, e_{i+1}, and e_{i+n}, such that e_{i+n} is BEFORE e_{i+1} and e_i is BEFORE e_{i+n}.

Retrograde
All events with lower subscripts are related by AFTER links to those with higher subscripts.

Syllepsis
There are clusters of events with contiguous subscripts where the clusters are distinguished by some non-temporal criterion, e.g., different clusters correspond to different places.

Zigzag
There are at least two clusters A and B of events, with events within each cluster related to each other by DURING or SIMULTANEOUS, with every event in cluster A being BEFORE any event in cluster B. The sequence of event subscripts often shows alternation between events in A and B.

To examine Zigzag in more detail, let us revisit the narratologically famous sentence in *Jean Santeuil* that we encountered first in Chap. 3 and in this chapter in Example 5.12. Zigzag is clearly apparent in the annotation shown in Fig. 5.2. It can be seen that the passing is SIMULTANEOUS with the sometimes, the remembering is DURING the passing, the bringing is DURING the rainy days, the not loving is BEFORE the savoring, etc.

Subordinated relations such as the relation between being taken to discover and Buendia's remembering, "remembered" and "the rainy days" and "remembered" and "then," and well as between "thought" and "some day" are all represented in TimeML by subordinating links, or SLINKs. These are shown with wavy arrows in Figs. 5.1 and 5.2. Note that in Fig. 5.2 the rainy days ($t2$) remembered by the narrator are not temporally related to the ones postulated by the narrator by "then" ($t3$). Such relations are not handled within the

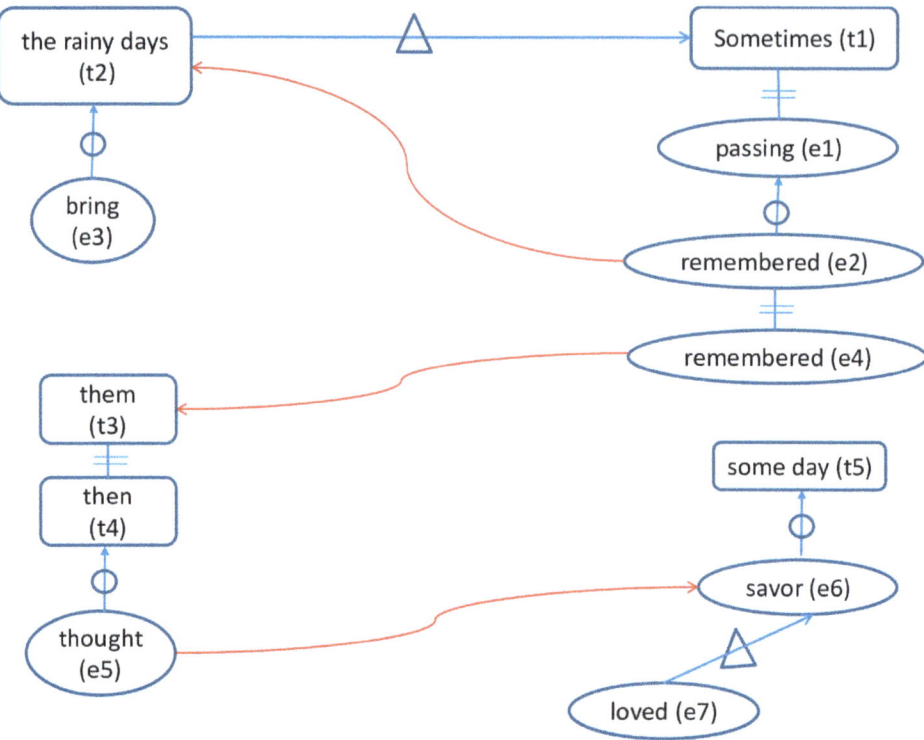

Fig. 5.2 Zigzagging in *Jean Santeuil*

strictly linear model of time that Allen's representation confines itself to, and there one has to resort to the branching time models mentioned above.

However, the annotation by humans of TLINKs in TimeML remains sparse, for several reasons:

- Inter-annotator agreement for TLINKs was only 0.55 F-measure, as reported by Mani et al. (2006) for the English TimeBank.[6] The poor agreement is partly because the Allen temporal relations are sometimes too fine-grained for the narrative at hand; in such cases, a disjunction of relations is preferred, but the guidelines, in the interests of simplicity, rule out marking disjunctions. As a result, one annotator may leave the relation out, while another may not. One way of addressing this is to allow for degrees of granularity in the representation, as argued by Mani (2005).

---

[6] Nevertheless, when a TLINK is posited by both annotators between the same pairs of events or times, the inter-annotator agreement on the labels goes up to 77% F-measure and 0.71 kappa, considered to be high agreement.

5.5 Temporal Representation

- Two annotators can arrive at equivalent but different temporal relations, e.g., one might declare A BEFORE B and B BEFORE C, while the other may also include A BEFORE C. This problem can be addressed by computing the transitive closure of both graphs, and comparing those closed graphs instead of the original annotations.
- Finally, given that the number of possible links is quadratic in the number of events and times, annotators tire easily; evidence from Bethard et al. (2007) suggests that many valid potential links in the TimeBank were missed in the annotation.

Given the need for addressing sparseness in the temporal relations annotated by humans, and the fact that in evaluating pairs of annotators or machines against annotations, transitive closures of the pair of temporal graphs were not compared, these earlier results underestimate the true difficulty of the problem. This has motivated a move towards far denser annotation, but with more coarse-grained relations. In the TimeBank-DENSE Corpus (Cassidy et al. 2014), almost every possible pair of events within a document is annotated with a temporal relation chosen from the set {*before, after, includes, is included, simultaneous, and vague*}. Transitive closures are computed as each temporal relation is annotated, forcing annotators to resolve inconsistencies. This results in more than ten times the number of temporal relations per document. Notably, *vague* constitutes about 46% of the relations, which might be viewed as an indication of the true difficulty of the annotation task. The inter-annotator agreement on this harder task tops off at .64 Kappa, which is lower than on the TimeBank. Overall, TimeBank-Dense is a maximalist approach, aiming at annotating all possible temporal relations.

A second corpus called MATRES (Ning et al. 2018) tries to enforce annotation scopes, called 'axes'. For instance, in Example 5.17, an event that is subordinated (as $e1$ is due to the subordinator 'tried') should not be related to one that is outside the scope of subordination, as is $e2$. However, $e0$ and $e1$ are within the same subordinated scope, so they should be related.

Example 5.17
Serbian police <u>tried</u> to eliminate$_{e0}$ the pro-independence Kosovo Liberation Army and restore$_{e1}$ order. At least 51 people were killed$_{e2}$ in clashes between Serb police and ethnic Albanians in the troubled region.

MATRES is rather minimalist, even going so-far as to exclude end-points of events when annotating temporal relations between them. With such a restricted annotation scheme, inter-annotator agreement goes up to 0.80 Kappa, which is much higher than reported elsewhere.

In future, more multilingual corpora along the lines of TimeBank-DENSE and MATRES are desirable. Pre-annotation by machines can play a key role here. Finally, synthetic corpora do have a role to play. The research of Fatemi et al. (2024) tests temporal reasoning of LLMs (ruling out the use of memorization) by generating synthetic graphs and questions and answers based on them.

5.6 Automatic Temporal Analysis and Generation

5.6.1 Linguistic Preliminaries

Consider the narrative understanding task of creating a chronology of events in a narrative. Given the many factors that can interact, inferring the underlying order of events from the discourse can be a challenge. The historical approaches used have included hand-constructed rules (e.g., Webber 1988, KampReyle1993, SchilderHabel2001, Lietal2005). Typical rules are the *narrative ordering rule* (past tense events succeed each other) and the *stative rule* (statives overlap with the event in the adjacent clause). For example, Lascarides and Asher (1993) provide both specific rules, e.g., that pushes normally cause falls, as well as general ones, such as causes preceding effects. The narrative ordering rule in particular seems a basic template that humans expect and is reflected in psychological experiments in human processing time.

However, such rules are often violated. In Example 5.18, from Webber (1988), the narrative rule does not work when the buying is viewed as an extended event; in that case, the picking out is DURING the buying.

Example 5.18
John bought Mary some flowers. He picked out three red roses.

Where is the knowledge to come from to handle a case like the latter? The answer, in part, is by machine-learning from annotated corpora.

5.6.2 Temporal Analysis

5.6.2.1 Time Expressions

Let us now turn to language understanding, where the problem is to extract the underlying temporal information from narrative discourse. The first extraction task we will consider is tagging the extents of time expressions (i.e., their offsets in the text) and resolving them. Given the availability of TIMEX2-annotated corpora, the most accurate approaches to TIMEX2 tagging have relied on machine learning from these corpora. In Task A of the TEMPEVAL-2 competition described by Verhagen et al. (2010), a single score was used for both tagging the extent and resolving the values of time expressions. The best systems scored 0.91 F-measure for Spanish and 0.86 F-measure for English. In general, one may conclude that tagging time expressions and resolving them is a problem that is being tackled quite well. Considering events, the Evita system of Sauri et al. (2005) tags TimeML events using particular morphological and syntactic patterns within a sentence (e.g., finite verbs, and event nouns). Such a system has a cited accuracy of about 80%. Tagging subordination relations via SLINKs, as in the case of Jean remembering the rainy days, or Aureliano Buendia remembering a distant afternoon, are in the 70% range, based on published results from the TARSQI system. Factuality profiles are tagged by Sauri and Pustejovsky (2012) at F-measures of 0.7–0.8.

5.6.2.2 Temporal Relations

We now dive into some of the computational history of automatically tagging temporal relations. It involves details of different computational methods, which can be skipped by those uninterested in or unfamiliar with such details.

> *Historical TLINK extraction**
> Previous research by myself and colleagues (Mani et al. 2006) has viewed labeling of temporal relations as a statistical classification problem: given an ordered pair of elements X and Y, where X and Y are events or times known to be related temporally via a TLINK, the classifier has to assign a single label from the set of TLINKs in TimeML. Using their TARSQI system, they obtained an accuracy of 59.68% for inter-event TLINKs and 82.47% for Event-Time TLINKs, when the training and test instances were not only different, but drawn from different documents in the TimeBank. Event ordering systems for narratives have been developed for a variety of languages, including English, Spanish, Swedish (Berglund et al. 2006), Chinese (Li et al. 2005), etc. Overall, in inter-event linking there is a ceiling effect, where accuracy scores do not rise much beyond the 0.7 F-measure.
>
> The statistical classifier TARSQI for TLINKs described above can produce an inconsistently annotated document. Since it merely inserts links between pairs of nodes A and B (events and/or times) in the graph, it does not take into account compositions with other nodes in the graph that have already been classified. One way of addressing this problem is to use a greedy method that first ranks the test instances by classifier confidence (i.e., preference) and then applies transitive closure axioms iteratively starting with the most preferred instances (Mani et al. 2007). A more general approach is to replace the second step with a method of combining preferences using closure axioms, but with the problem of consistency framed as a global optimization problem. Integer Linear Programming (ILP) is a machine learning framework that provides a generic method for solving the global optimization problem, while allowing for declarative (rule-based) specification of closure axioms. ILP thus guarantees that any solution found will be consistent. Working on medical case history summaries from the *New England Journal of Medicine*, (Bramsen et al. 2006) have developed a method that computes the temporal ordering of elements that are temporal segments corresponding to distinct time frames, each of which can contain one or more events. For example, a medical narrative might have segments on a patient's admission, a previous hospital visit, and the onset of symptoms. To infer temporal relations, their use of ILP resulted in an accuracy of 84.3%.
>
> An alternative to ILP is the use of Markov Logic Networks, as used in Yoshikawa et al. (2009). Here the closure constraints do not have to be all-or-nothing ones as in ILP, allowing instead for *soft* ones such as the following:

Property 5.1 If time $t1$ is BEFORE the Document Creation Time (or DCT, i.e., document date) and event $e1$ OVERLAPS DCT, then $e1$ is AFTER $t1$.

Using Markov Logic Networks, weights for such rules can be learned from the training data. Such an approach yielded the best results on TEMPEVAL-1 data for Task A (events and times in the same sentence) as well as Task C (events in two consecutive sentences).

Contemporary approaches
Contemporary approaches to temporal relation extraction rely on training from annotated data using neural methods. Research by Huang et al. (2023) converts the interval-based representation of events to a point-based one, resulting of course in only three possible relation between points being compared. Logical interference in the point-based representation is then carried out, within a Transformer architecture. The model achieves F-measures of 0.71 on TimeBank-Dense and 0.84 on MATRES, outperforming the previous state-of-the-art by 0.3%.

An especially interesting approach by Zhang et al. (202) is to use syntactic information in the form of extracted syntactic dependency graphs to help guide temporal relation extraction. Given a sentence, event detection is carried out conventionally using the BERT Transformer, and then a Graph Transformer is used which makes use of the syntactic dependency graph of the sentence. The latter considers in its Self-attention mechanism the triples each event mention is involved with, e.g., in the sentence "He won the Gusher Marathon, finishing in 3:07:35", 'won' will be in the nominal-subject relation to 'He' and a direct object relation to 'Marathon', and 'finishing' as an adverbial clause relation to 'won', etc. Finally, the two event mention representations are concatenated to predict their temporal relationship. The model has F-measures of 0.83 on MATRES and 0.74 on TB-Dense, surpassing the previous state-of-the-art by up to 7.9%.

5.6.2.3 LLMs Detecting Genette's Narrative Orderings
How well can modern LLM approaches interpret the narrative orderings distinguished by narratology? To make it more readable, I print the GPT-4o orderings for the text below each sentence. To save space, I don't include the obvious user prompts.

- Analepsis:

Example 5.19
We went$_{e1}$ along the fence and came$_{e2}$ to the garden fence, where our shadows were.
BEFORE(e1, e2)
My shadow was higher than Luster's on the fence. We came$_{e3}$ to the broken place and went$_{e4}$ through it.
BEFORE(e2, e3) & SIMULTANEOUS(e4, e3)
"Wait a minute," Luster said$_{e5}$. "You snagged$_{e6}$ on that nail again. Can't you crawl$_{e7}$ through

5.6 Automatic Temporal Analysis and Generation

here without snagging$_{e8}$ on that nail?"
BEFORE(e6, e5) & SIMULTANEOUS(e7, e6) & SIMULTANEOUS(e8, e7)
Caddy uncaught$_{e9}$ me and we crawled$_{e10}$ through.
BEFORE(e9, e4) & SIMULTANEOUS(e10, e9)
Uncle Maury said$_{e11}$ to not let anybody see$_{e12}$ us, so we better stoop$_{e13}$ over, Caddy said$_{e14}$.
BEFORE(e10, e11)
"Stoop over, Benjy. Like this, see." We stooped$_{e15}$ over and crossed$_{e16}$ the garden
BEFORE(e13, e12) & SIMULTANEOUS(e15, e13) & BEFORE(e15, e16)
where the flowers rasped$_{e17}$ and rattled$_{e18}$ against us.
SIMULTANEOUS(e17, e16) & SIMULTANEOUS(e18, e17)
The ground was hard.

The tags happen to be correct, including the crucial one marking the transition (shown in boldface), but GPT-4o fails to detect any flashback, in short, it couldn't segment the subtle transition to Faulkner's childhood scene.

- Prolepsis:

Example 5.20
At that very first high-school date she had$_{e1}$ a sudden image of the two of them sitting$_{e2}$ together in their old age in rocking chairs on the front porch of her grandparents' home.
MODAL(e1, e2)
She knew$_{e3}$ then that he was$_{e4}$ the right choice.
BEFORE(e1, e3) & EVIDENTIAL(e3, e4)

The flash-forward is handled correctly. However, there is a small error. The MODAL relation is a subordination relation brought by modal verbs like 'promise', 'want', 'know', etc. The EVIDENTIAL relation is another subordination relation introduced by reporting or perception verbs, e.g., 'say', 'see', 'having an image', etc. The LLM has swapped the pair.

- Retrograde:

Example 5.21
Tina was touched$_{e1}$ that her boyfriend had pretended$_{e2}$ that the spaghetti she had cooked$_{e3}$ for him was good, to spare her feelings.
EVIDENTIAL(e1, e2) & EVIDENTIAL(e2, e3) & BEFORE(e3, e2)
She had tried$_{e4}$ it herself and realized$_{e5}$ it was disgusting after he had said$_{e6}$ it was good, upon eating$_{e7}$ the whole plate.
BEFORE(e3, e4) & BEFORE(e4, e5) & BEFORE(e6, e5)
& SIMULTANEOUS(e2, e6)
& BEFORE(e7, e2) & BEFORE(e7, e6) & BEFORE(e7, e4)
& BEFORE(e7, e5)

Retrograde is reasonably challenging and is handled fine.

- Syllepsis:

Example 5.22
The gunshot was drowned$_{e1}$ by the sound of the midnight train$_{t1:TNI}$ pulling$_{e2}$ away.
SIMULTANEOUS(e1, e2)
In the hotel room, she hesitated$_{e3}$ as she took$_{e4}$ the revolver out of her suitcase.
BEFORE(e4, e3)
In the restaurant, he promised$_{e5}$ her they would be together forever once they reached$_{e6}$ Venice.
BEFORE(e5, e6) & MODAL(e5, e6)

Here the system is careful to annotate with UNKNOWN tags the links between events it's not sure of, handling syllepsis correctly.

- Zigzag:

Example 5.23
Sometimes passing$_{e1}$ in front of the hotel he remembered$_{e2}$ the rainy days when$_{t1:PTXD}$ he used to bring$_{e3}$ his nursemaid that far, on a pilgrimage.
BEFORE(e3, e2) & BEFORE(e3, t1)
But he remembered$_{e4}$ them without the melancholy that he then thought$_{e5}$ he would surely some day$_{t2:FutureRef}$ savor$_{e6}$ on feeling$_{e7}$ that he no longer loved$_{e8}$ her.
SIMULTANEOUS(e4, e2) & MODAL(e5, e6) & MODAL(e6, e7)
& EVIDENTIAL(e7, e8) & BEFORE(e8, e7) & BEFORE(e5, t2)

This too is correct though still incomplete.

The results on these narrative orderings seem satisfactory, and perhaps, if evaluated formally on a corpus, comparable or beyond what humans (including literary critics) could do if required.

5.6.2.4 Scene-Setting Narratives

Scene-setting is basic to any narrative. Consider Balzac's novella *Sarrasine*, which has been of interest to narratologists ever since Barthes (1975) analyzed it in structuralist-semiotic terms. The opening sentence and GPT-4o annotation is shown in Example 5.24.

Example 5.24
I was buried$_{e1}$ in one of those profound reveries$_{e2}$ to which everybody, even a frivolous man, is subject in the midst of the most uproarious festivities$_{e3}$.

5.6 Automatic Temporal Analysis and Generation

EVIDENTIAL(e1, e2) & MODAL(e2, e3)
& SIMULTANEOUS(e2, e3)
The clock on the Elysee-Bourbon had just struck$_{e4}$ midnight$_{t1}$
BEFORE(e4, t1) & SIMULTANEOUS(e3, e4)

Here the state of being subject to reveries ($e2$) is viewed as being SIMULTANEOUS with the festivities ($e3$), whereas it should be DURING. Both are, however, habitual events, which aren't modeled appropriately in TimeML. The festivities ($e3$) should also not be SIMULTANEOUS with the clock's striking ($e4$). Claude 3.5 Sonnet and Gemini Advanced are even worse. It is sad to see that in the 18 years since our statistical classifier for TLINKs (trained on annotated data) was published, the Transformer-based tools don't fare much better.

Another instance of scene-setting descriptions in *Sarrasine* is in Example 5.25.

Example 5.25
The trees, being partly covered$_{e1}$ with snow, were outlined$_{e2}$ indistinctly against the grayish background formed$_{e2}$ by a cloudy sky, barely whitened$_{e4}$ by the moon.

Here the temporal relations between the events should, arguably, all be OVERLAPs. GPT-4o treats them all as SIMULTANEOUS, which may be marginally correct.

5.6.2.5 LLMs Detecting Metalepsis

Let us now reconsider from a temporal standpoint the metalepsis that we encountered in Cortazar's *The Continuity of Parks* (Cortazar 1967), discussed in Chap. 3, Example 3.3.3. There the character (call him A) is reading$_{e1}$ a novel involving a character B who eventually sneaks up on A to kill$_{e2}$ B while the latter is reading$_{e3}$. Here the subordinated discourse introduced by the reading event extends into the embedding story. In our annotation scheme, there is no cycle in time here or inconsistency, because there is an SLINK from $e1$ to $e2$ and to all the embedded events in the embedded story, including $e3$.

GPT-4o will happily ignore metalepsis unless reminded of it, in which case it will attempt to find it, as in examples like the following, where the italicized input is from the Gemini Advanced output we saw in Chap. 3, Example 3.17:

Example 5.26

> **User Prompt**: Generate the NarrativeML for this story. Show me only the part involving metalepsis, using an SLINK. Make sure your output is short but still comprehensible.
> *There was a power outage in the Smith residence. Fred could barely see in front of him.*

> *He crawled around in the dark, feeling his way around. Finally, he felt his hand wrap around a familiar item. He turned on the flashlight and illuminated the author, who was frantically scribbling the final words of the story by the light of his laptop screen.*
> **Machine:**
> ```
> <EVENT id="e5" type="ACTION"
>     participants="c1 c3">turn_on_flashlight</EVENT>
> <EVENT id="e6" type="ACTION" participants="c1 c2">
>     illuminate_author</EVENT>
> <SLINK id="s1" type="MODAL" eventID="e6"
>     subordinatedEventID="e5"/>
> ```
> Fred crosses the narrative boundary to illuminate the author.
>
> This excerpt captures the metaleptic interaction where Fred, within the story, illuminates the author, crossing the narrative boundary. The MODAL SLINK signifies the hypothetical or boundary-crossing nature of this event.

5.6.2.6 Temporal Graph Decomposition

Decomposing the graph of temporal relations reveals key components, such as different stages of a narrative, or when a given *character* appears. Doing the latter not only simplifies the temporal graph, but allows for comparison of character behaviors. For example, consider the *The Travelers and the Bear* from Aesop's Fable, shown in Example 5.27. The temporal graph for it is annotated as shown in Fig. 5.3, redrawn from Bethard et al. (2012, p. 2722).[7] Here the arrows with ovals, as before, indicate DURING relations, while the other arrows indicate BEFORE relations. The subgraph where the nimbler traveler disappears from the narrative is indicated by a dashed rectangle. In contrast, the portions of the narrative allocated to the nimbler traveler's presence are indicated by dashed teardrops. It is clear that the nimbler traveler is backgrounded, and that the other traveler not only gets more press (in keeping with his exemplary behavior), but occupies a larger contiguous stretch of the timeline.

Example 5.27

Two Travelers were on the road together, when a Bear suddenly appeared on the scene. Before he observed them, one made for a tree at the side of the road, and climbed up into the branches and hid there. The other was not so nimble as his companion; and, as he could not escape, he threw himself on the ground and pretended to be dead. The Bear came up and sniffed all round him, but he kept perfectly still and held his breath: for they say that a bear will not touch a dead body. The Bear took him for a corpse, and went away. When the coast was clear, the Traveler in the tree came down, and asked the other what it was the Bear had whispered to him when he put his mouth to his ear. The other replied, "He told me never again to travel with a friend who deserts you at the first sign of danger."

---

[7] For automatic annotation of such stories, see Kolomiyets et al. (2012).

5.7 Duration and Tempo

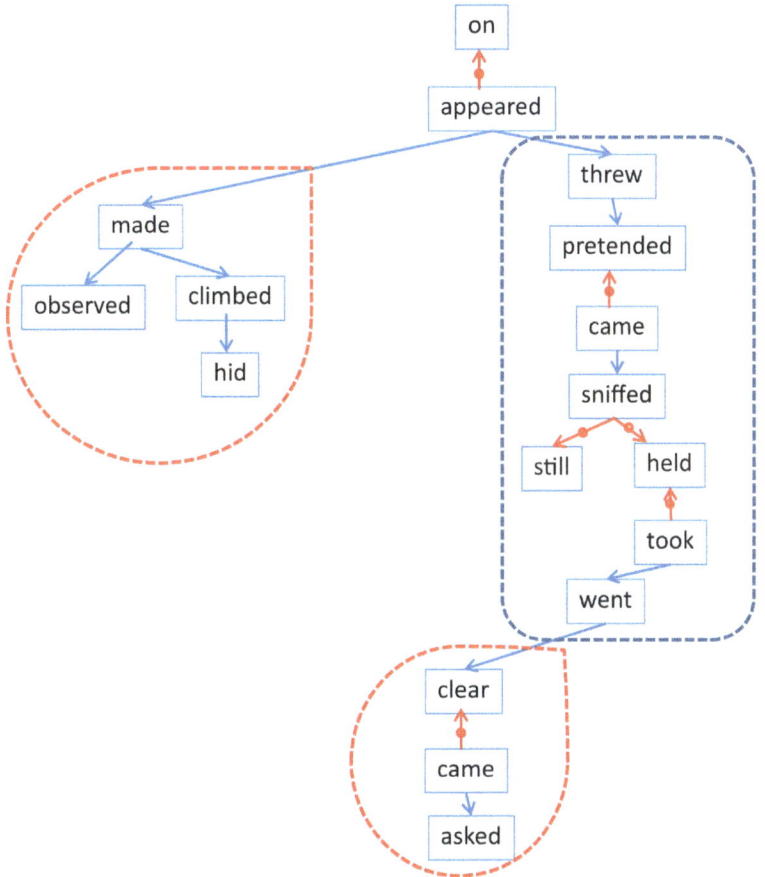

Fig. 5.3 Temporal graph for *The Travelers and the Bear*

5.7 Duration and Tempo

5.7.1 Duration

Humans have strong intuitions about how long particular events last (e.g., an invasion lasts longer than a sneeze). In some cases, the narrative itself will indicate how long particular events last, but in many cases it is left implicit. it is possible to add to the markup estimates of the minimum and maximum bounds for events, as Pan et al. (2011) have done for TimeML (so as to cover 80% of the probable scenarios given the text context). Humans tend to agree almost 90% of the time on such bounds. However, there is still some delicacy involved in the judgments for the more complex task of setting more fine-grained bounds than simply a day or longer; agreement was around 75% for these.

Here is an example annotation, which records a saying event as taking between 5 s and 5 min and the single event of visiting of a site as taking between 10 min and 1 day:

Example 5.28

```
The official <EVENT min='PT5S' max='PT5M'>said</EVENT> the
site could only be <EVENT min='PT10M'
max='P1D'>visited</EVENT> by a special team of U.N.
monitors and diplomats.
```

An automatic tagger (Pan et al. 2011) trained on the TimeBank sub-corpus (containing 2330 events thus annotated) scored 76% accuracy in determining whether an event lasted a day or longer (ibid.).

A related issue is determining how much time has elapsed during an entire passage. A study by Underwood (Underwood 2018) found that the tendency towards breathless minute-by-minute description has shown a steady increase since the eighteenth century. That study was carried out manually by three people, focusing on 250-word passages. To go from durations of individual events (with or without temporal bounds) to those of event sequences in a passage requires a substantial degree of inference. Subsequently, a system by Yauney et al. (2021) was able to automate the analysis using a bag-of-words machine-learning model (one that creates an array of frequencies of corpus words in each passage). An intriguing exploration by Underwood (2023) revisits this issue of elapsed times by using Chain-of-Thought prompting with GPT-4, providing four example passages with elapsed times and the reasoning indicated. The system is instructed to guess in the absence of explicit references to time, and if there is a range of possible times, to take the midpoint of the range. The results agreed reasonably well with the judgments of human readers (Pearsons r correlation of 0.68), and was much better than the bag-of-words model of Yauney et al. (2021), showing the sophistication of Transformer-based methods. The reliability is still less than human agreement (Pearsons r correlation of 0.74), but it is of course much, much faster than humans, which is one of the wins of this sort of automation.

How the elapsed time changes in a text is the basis for pacing or tempo, discussed next. In doing so, we revisit the issue of event durations from the perspective of commonsense knowledge.

5.7.2 Tempo

Fiction thrives on variations in tempo. A text may pay microscopic attention to details occupying small time intervals, or it may zoom out and briskly cover, in a few sentences, entire decades.

Narratives use techniques such as stretching the discourse time to linger on a particular moment (as in *time-slice* or *bullet time* movies), or devoting exorbitant amounts of

5.7 Duration and Tempo

time to relatively inconsequential events for satirical or voyeuristic effects (as can be seen, respectively, in Voltaire's *Candide* and Robbe-Grillet's *Jealousy*), or providing speeded-up descriptions, as in James Michener's *Hawaii*, cited by Jahn (2005):

Example 5.29
The years passed. The sun swept through its majestic cycles. The moon waxed and waned, and tides rushed back and forth across the surface of the world.

> **More on elapsed time**
> Events that are simple time-fillers or of no use other than being required for purposes of moving the narrative along or providing information, what are known in narratology (1977) as *indicial actions*, can be compressed in various ways in the text. However, non-indicial events which are of long durations which are allocated little reading length, and vice versa, are likely to have more *surprisal*. That is what happens, for example, when there is a substantial temporal gap, as in the case of novels that catch up with events after a lapse of several decades. This lapse requires that the reader try to imagine what might have happened to characters in the interim.

In classical narratology, Genette (1980) classifies tempos into three kinds: isochronous, accelerated, and decelerated. The temporal relations being considered in our temporal graphs are more expressive than merely precedence and equality in Genette's narratology, since we allow for time intervals which are related in the seven different ways. Discourse time itself is not directly captured in NarrativeML, though this is a matter of detail; the indices ($e3$, etc.) might be easily extended to include a measure of offset into the narrative (a similar point is also made by Meister (2005)). Once that is added, the *story time* of a narrative (the time it occupies in the timeline) can be compared to its *discourse time*, i.e., the length of the text in terms of the time taken to read it. That, in turn, can if desired be normalized to a function of the number of words used to recount the event (e.g., using an average reading rate, such as 300 words per minute). The ratio of story time to discourse time, offers a computational measure of narrative pace.

Cases where inconsequential events are dealt with at great length, or where significant events are passed over very speedily, can illustrate interesting narrative effects.

The literary scholar (2001) cites the following passage from Voltaire's *Candide* to illustrate some interesting variations in tempo. This passage occurs immediately after one when the wise men of Portugal decide, after most of Lisbon is destroyed by the 1755 earthquake, that giving the people a beautiful auto-da-fe will prevent further tremors. (Trigger WARNING: Voltaire's satirical style is not for the faint-hearted or the easily offended).

Example 5.30

They were <u>conducted</u> to separate apartments, extremely cold, as they were never incommoded by the sun. Eight days after, they were <u>dressed</u> in san-benitos and their heads <u>ornamented</u> with paper mitres. The mitre and san-benito belonging to Candide were <u>painted</u> with reversed flames and with devils that had neither tails nor claws; but Pangloss's devils had claws and tails and the flames were upright. They <u>marched</u> in procession thus habited and <u>heard</u> a very pathetic sermon, followed by fine church music. Candide was <u>whipped</u> in cadence while they were <u>singing</u>; the Biscayner, and the two men who had <u>refused</u> to <u>eat</u> bacon, were <u>burnt</u>; and Pangloss was <u>hanged</u>, though that was not the custom. The same day the earth <u>sustained</u> a most violent <u>concussion</u>.

The speed is quick until the description of Candide and Pangloss's costumes, and then the pace again picks in the last two sentences. The speedy and offhand account of the hanging of Pangloss, who is a principal character only enhances Voltaire's characteristic satirical (and, sadly, somewhat prejudiced) treatment.

Once can view the timeline of this passage as follows. Here RT is the reference time, and *t-conduct* is the duration of being conducted to separate apartments.

Example 5.31

..
$t1 = RT + t\text{-conduct}$
$t2 = t1 + P8D$
$t3 = t2 + t\text{-dress}$
$t4 = t3 + t\text{-march}$
$t5 = t4 + t\text{-sermon}$
$t6 = t5 + t\text{-whip}$
$t7 = t6 + t\text{-hang}$
$t8 = t6 + t\text{-burn}$

I have annotated the *Candide* passage in accordance with the guidelines from Pan et al. (2011). Figure 5.4 shows a bar chart of the story time (length of each event's annotated durations) (in the chart legend, story time is called STORY) as a percentage of the sum of those lengths.[8] The discourse time devoted to each event (length in words, shown in the legend as DISCOURSE) is also shown as a percentage of the sum of those lengths.[9]

Comparing the two peaks, we see that while 44% of the story time is occupied by the sermon (a mere 8% of the description length), almost 60% of the description is spent on the dressing (which is no more than 14% of the simulation time). The discourse time is also out of phase with the story time.

---

[8] From Mani (2010).

[9] Since nothing much happened during the 8-day period, we focus on the passage beginning at the dressing.

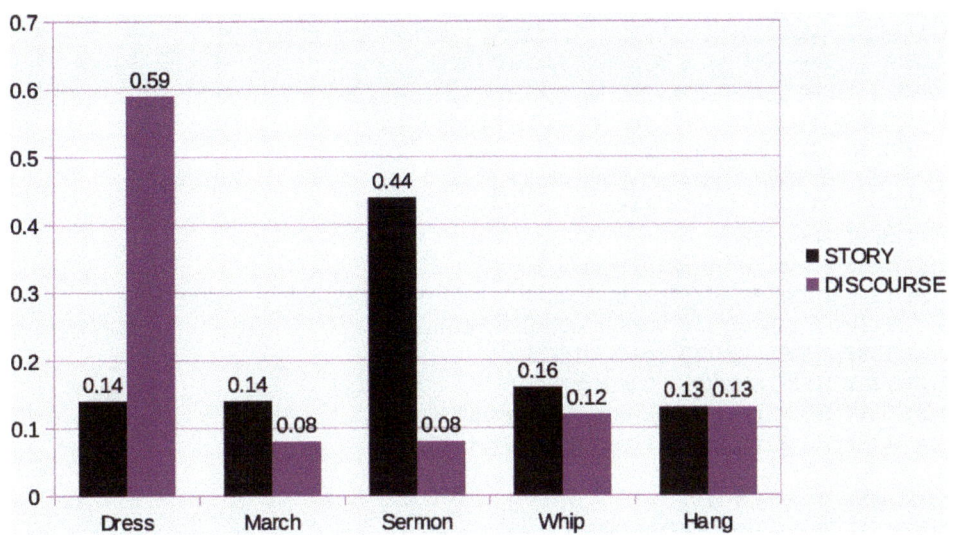

Fig. 5.4 Discourse time versus story time for *Candide* passage

It is instructive to compare my times for 'conduct' (5–15 min), 'march' (5 min to 1 h), and 'burn' (5 min to 1 h) in the *Candide* passage to the times for similarly-named events in Pan et al. (2011)'s duration-annotated TimeBank. There, we find the much longer 1–2 months for the Khmer Rouge conducting a search for a body, 1–2 years for conducting auction business, both of which involve the rather different sense of carrying out an activity rather than leading someone somewhere; the somewhat longer 1–3 h for a march (in the related sense of a demonstration) by Catholics in Belfast, and 30 min to 6 h for one by ethnic Albanians in Turkey, and the shorter time of 30 s to 5 min for those marching Albanians to burn a Serbian flag. The latter event involves a time interval that is comparable to a subinterval of the incineration of the poor Candide. Obviously, judgments of duration can only be compared for the same type of event (i.e., a disambiguated and similar word sense or embedding) in a similar occurrence context.

5.8 NarrativeML with Temporal Information and LLM Annotation

We can now extend NarrativeML to include temporal and subordination information. This is shown in Example 5.29.

Example 5.32

```
<?xml version="1.0" encoding="UTF-8"?>
<!DOCTYPE NarrativeML [
...
```

```
<!ELEMENT NARRATIVE (#PCDATA | NARRATOR | AUDIENCE | CHARACTER |
EVENT | TIME | TLINK | SLINK | EVALUATION |
GOAL | TEMPO | NEC | PLOT | SEGMENT)*>
...
  <ATTLIST TIME beginPoint IDREF #IMPLIED>
   <!ATTLIST TIME endPoint IDREF #IMPLIED>
...
<!ELEMENT TLINK EMPTY>
<!ATTLIST TLINK id ID #REQUIRED>
<!ATTLIST TLINK type (SIMULTANEOUS | BEFORE | IMM_BEFORE | OVERLAPS
  | BEGINS | ENDS | INCLUDES | EQUALS) #REQUIRED>
<!ATTLIST TLINK eventID IDREF #IMPLIED>
<!ATTLIST TLINK timeID IDREF #IMPLIED>
<!ATTLIST TLINK relatedToEvent IDREF #IMPLIED>
<!ATTLIST TLINK relatedToTime IDREF #IMPLIED>
...
<!ELEMENT SLINK EMPTY>
<!ATTLIST SLINK id ID #REQUIRED>
<!ATTLIST SLINK eventID  IDREF #REQUIRED>
<!ATTLIST SLINK subordinatedEventID  IDREF #REQUIRED>
...
]>
```

Example 5.33 shows the opening sentence of a travel blog and its machine annotation by GPT-4o in NarrativeML, including plot-related NECs. This is a one-shot example where the net was first given the NarrativeML DTD and an example annotation. The blog is drawn from Mani and Pustejovsky (2012).

Example 5.33

> **User Prompt**: Generate the NarrativeML for this text:
> *March 7, 2006. Leaving San Cristobal de las Casas, I biked with Gregg and Brooks for one more day.*
> **Machine**:
> ```
> <NarrativeML version="0.4">
>   <NARRATIVE id="n1" title="Biking with Friends"
>   medium="descriptive_text">
>     <NARRATOR id="n2" type="present" coref="c1" exists="true"
>               order="CHRONICLE" distance="DIRECT"
>               perspective="INTERNALLY_FOCALIZED"
>               timeRelation="SUBSEQUENT"/>
>     <CHARACTER id="c1" name="I" type="animate" exists="true"
>     accessibleTo="c1"/>
>     <CHARACTER id="c2" name="Gregg" type="animate" exists="true"
> ```

```xml
    accessibleTo="c1"/>
  <CHARACTER id="c3" name="Brooks" type="animate" exists="true"
    accessibleTo="c1"/>
  <CHARACTER id="c4" name="San_Cristobal_de_las_Casas"
    type="inanimate" exists="true" accessibleTo="c1"/>
  <CHARACTER id="c5" name="Bicycle" type="inanimate" exists="true"
    accessibleTo="c1"/>
  <SEGMENT id="s1" title="Leaving San Cristobal and Biking">
    <TIME id="t1" value="March 7, 2006"/>
    <EVENT id="e1" type="ACTION" participants="c1 c2 c3 c4"
      duration="PT1H">leave_San_Cristobal</EVENT>
    <TLINK id="tr1" type="IS_INCLUDED" eventID="e1" timeID="t1"/>
    <EVENT id="e2" type="ACTION" participants="c1 c2 c3 c5"
      duration="PT1D">bike_together</EVENT>
    <TLINK id="tr2" type="SIMULTANEOUS" eventID="e2"
      relatedToEvent="e1"/>
  </SEGMENT>
  <TEMPO storyTime="PT1D" readingLength="PT30S"/>
  <NEC id="nec1" entity="c1" events="e1 e2"/>
  <NEC id="nec2" entity="c2" events="e1 e2"/>
  <NEC id="nec3" entity="c3" events="e1 e2"/>
  <NEC id="nec4" entity="c4" events="e1"/>
  <NEC id="nec5" entity="c5" events="e2"/>
</NARRATIVE>
</NarrativeML>
```

It can be seen that the system fails to identify San Cristobal de las Casas as a place, and doesn't properly delimit the extents of event mentions. The aspectual event of "leaving" is annotated with varying durations in the TimeBank extensions of Pan et al. (2011), with the maximum lower bound being one month. The context in our case is clearly different. For this blog, I would estimate an average duration of a generous two hours to get out of town, allowing for traffic, but the machine guesses one hour. The event of "biking" is not annotated in their TimeBank; I would estimate an average duration estimate of nine hours; the machine guesses one day. For the tempo, the 19-word one-sentence micro-story spanning a story time of one day is estimated, given a reading speed of 300 words per minute, as taking approximately four seconds to read; the machine seems to think we humans are slow, and that we instead need a full thirty seconds to read it (thirty seconds is not that short a duration for us: it can seem like forever when you're stuck in a lift!). For a far more detailed temporal analysis by humans of the longer travel blog from which this excerpt is taken, see Mani and Pustejovsky (2012, pp. 130–133).

5.9 Narratological Implications and Conclusion

The creation of temporal graphs for narratives makes possible new corpus-based lines of investigation that are relevant to the interests of humanities narratology. Here are some examples:

- Finding narratives whose temporal graphs illustrate a pattern. Patterns such as the Chronicle or Zigzag of Genette (1980) are already trivially computable for any TimeML-annotated corpus, as suggested by the examples in the previous section. Patterns for cases where subordinated events are involved, modeled with branching time, present an area ripe for further narratological annotation.
- Identifying changes of tempo in narratives.
- Decomposing temporal relation graphs to identify particular components, such as different segments of a story, when a given *character* appears, and comparisons of character behavior.

In this chapter, I have described in some detail methods for representing and computing events, times, and temporal relations in narrative, the latter represented in terms of chronologies that are partial orderings based on qualitative relations from the interval calculus. Quantitative (or metric) constraints are also represented, along with commonsense durations for events. Phenomena like tempo are also characterized and annotated. The representation impinges on issues of modality, addressed partially through subordination relations (represented in NarrativeML) and event factuality representations (represented in a separate annotation scheme from Sauri and Pustejovsk 2012). These representations add an important empirical dimension to corpus-based narratology, allowing for particular temporal patterns in corpora to uncover, as we have shown, novel narratological insights.

As I have shown, AI systems are able to extract and resolve time expressions successfully and order events in narrative discourse in terms of an underlying partial order, and even represent all the different orderings discussed by narratologists. For more theoretical details on how these computational approaches relate to narratological and literary discussions of time, see Mani (2010).

However, many interesting research challenges remain:

- The modal issues mentioned above need further formal modeling beyond the interval calculus, including the use of models of branching time.
- The inferences regarding commonsense durations of events are highly coarse-grained and need to be extended much further.
- There is a pressing need to extend annotation schemes like TimeML to address phenomena characteristic of narrative such as habituals and scene-setting narratives.

References

Allen, James F. (1983). "Maintaining knowledge about temporal intervals". In: *Communications of the ACM* 26.11, pp. 832–843. URL: https://doi.org/10.1145/182.358434.

Allen, James F. (1984). "Towards a general theory of action and time". In: *Artificial Intelligence* 23.2, pp. 123–154. URL: https://doi.org/10.1016/0004-3702(84)%2090008-0.

Allen, James F. and Patrick J. Hayes (1985). "A common-sense theory of time". In: *Proceedings of the Ninth International Joint Conference on Artificial Intelligence (IJCAI 1985)*. Los Angeles, CA, pp. 528–531. URL: https://www.ijcai.org/Proceedings/85-1/Papers/101.pdf.

Andre, Bittar, Pascal Amsili, Pascal Denis, and Laurence Danlos (2011). "French TimeBank: an ISO-TimeML annotated reference". In: *Proceedings of the 49th Annual Meeting of the Association for Computational Linguistics: Human Language Technologies (ACL-HLT'2011), Portland Oregon*, pp. 130–134. URL: https://aclanthology.org/P11-2023/.

Barthes, Roland (1975). *S/Z: An Essay*. New York: Hill and Wang.

Barthes, Roland (1977). *Image, Music, Text*. New York: Hill and Wang.

Berglund, Anders, Richard Johansson, and Pierre Nugues (2006). "A machine learning approach to extract temporal information from texts in Swedish and generate animated 3D scenes". In: *Proceedings of Eleventh Conference of the European Chapter of the Association for Computational Linguistics (EACL-2006)*. Trento, Italy, pp. 385–392. URL: https://aclanthology.org/E06-1049.pdf.

Bethard, Steven, James H. Martin, and Sara Klingenstein (2007). "Finding temporal structure in text: machine learning of syntactic temporal relations". In: *International Journal of Semantic Computing* 1.4, pp. 441–457. URL: https://doi.org/10.1142/S1793351X07000238.

Bethard, Steven, Oleksandr Kolomiyets, and Marie-Francine Moens (2012). "Annotating story timelines as temporal dependency structures". In: *Proceedings of the 8th International Conference on Language Resources and Evaluation (LREC'2012)*. Istanbul, Turkey, pp. 2721–2726. URL: https://aclanthology.org/L12-1182/.

Bramsen, P., P. Deshpande, Y. K. Lee, and R. Barzilay (2006). "Inducing temporal graphs". In: *Proceedings of the Conference on Empirical Methods in Natural Language Processing (EMNLP-2006)*. Sydney, Australia, pp. 189–198. URL: https://doi.org/10.3115/1610075.1610105.

Bunt, Harry and Chwhynny Overbeeke (2008). "An extensible compositional semantics for temporal annotation". In: *Proceedings of the Second Linguistic Annotation Workshop*. Marrakech, Morocco. URL: https://let.uvt.nl/general/people/bunt/docs/temsem.pdf.

Caselli, Tommaso, Valentina Bartalesi Lenzi, Rachele Sprugnoli, Emanuele Pianta, and Irina Prodanof (2011). "Annotating events, temporal expressions and relations in Italian: the It-TimeML experience for the Ita-TimeBank". In: *Proceedings of the Fifth Linguistic Annotation Workshop (2011)*. Portland, Oregon, pp. 143–151. URL: https://aclanthology.org/W11-0418/.

Cassidy, Taylor, Bill McDowell, Nathanael Chambers, and Steven Bethard (2014). "An Annotation Framework for Dense Event Ordering". In: *Proceedings of the 52nd Annual Meeting of the Association for Computational Linguistics*, pp. 501–506. URL: https://aclanthology.org/P14-2082.pdf.

Chatman, Seymour (1980). *Story and Discourse: Narrative Structure in Fiction and Film*. Ithaca: Cornell University Press.

Comrie, Bernard (1986). *Tense*. Cambridge, UK: Cambridge University Press.

Cortazar, Julio (1967). *End of the Game and Other Stories*. New York: Pantheon Books. URL: https://biblioklept.org/2023/06/09/a-continuity-of-parks-julio-cortazar/.

Costa, Francisco and Antonio Branco (2012). "TimeBankPT: a TimeML annotated corpus of Portuguese." In: *Proceedings of the 8th International Conference on Language Resources and Evaluation (LREC'2012)*. Istanbul, Turkey, pp. 3737–3734. URL: https://aclanthology.org/L12-1096/.

Fatemi, Bahare et al. (2024). *Test of Time: A Benchmark for Evaluating LLMs on Temporal Reasoning.* URL: https://arxiv.org/pdf/2406.09170.

Ferro, Lisa, Laurie Gerber, Inderjeet Mani, Beth Sundheim, and George Wilson (2005). *TIDES 2005 standard for the annotation of temporal expressions.* Tech. rep. McLean, VA: MITRE. URL: http://timex2.mitre.org/annotation%5C_guidelines/timex2%5C_annotation%5C_guidelines.html.

Forascu, Corina, Radu Ion, and Dan Tufis (2007). "Semi-automatic annotation of the Romanian TimeBank 1.2." In: *Proceedings of the RANLP 2007 Workshop on Computer-aided language processing (2007).* Borovets, Bulgaria.

Genette, Gerard (1980). *Narrative Discourse.* Ithaca: Cornell University Press.

Graesser, Arthur C., Brent Olde, and Bianca Klettke (2002). "How does the mind construct and represent stories?" In: *Narrative Impact: Social and Cognitive Foundations.* Ed. by Melanie Green, Jeffrey Strange, and Timothy Brock. New Jersey: Lawrence Erlbaum.

Grall, Clare, Josefa Equita, and Emily S Finn (2023). "Neural unscrambling of temporal information during a nonlinear narrative". In: *Cerebral Cortex* 33.11, pp. 7001–7014. URL: https://doi.org/10.1093/cercor/bhad015.

Huang, Quzhe, Yutong Hu, Shengqi Zhu, and Yansong Feng (2023). "More than classification: A unified framework for event temporal relation extraction". In: *Proceedings of the 61st Annual Meeting of the Association for Computational Linguistics,* pp. 9631–9646. URL: https://arxiv.org/pdf/2305.17607.

Ireland, Ken (2001). *The Sequential Dynamics of Narrative.* Madison: Fairleigh Dickinson University Press.

Jahn, Manfred (2005). *Narratology: A Guide to the Theory of Narrative.* English Department, University of Cologne. URL: http://www.uni-koeln.de/~ame02/ppp.htm.

Kautz, Henry A. and Peter B. Ladkin (1991). "Integrating metric and qualitative temporal reasoning". In: *Proceedings of the Ninth National Conference on Artificial Intelligence (AAAI'91).* Washington, DC: AAAI Press, pp. 241–246. URL: https://cdn.aaai.org/AAAI/1991/AAAI91-038.pdf.

Kolomiyets, Oleksandr, Steven Bethard, and Marie-Francine Moens (2012). "Extracting narrative timelines as temporal dependency structures". In: *Proceedings of the 50th Annual Meeting of the Association for Computational Linguistics (ACL'2012).* Jeju, Korea, pp. 88–97. URL: https://aclanthology.org/P12-1010/.

Lascarides, Alex and Nicholas Asher (1993). "Temporal relations, discourse structure and commonsense entailment". In: *Linguistics and Philosophy* 16, pp. 437–494. URL: https://doi.org/10.1007/BF00986208.

Li, Wenjie, Kam-Fai Wong, and Chunfa Yuan (2005). "A model for processing temporal references in Chinese". In: *The Language of Time: A Reader.* Ed. by Inderjeet Mani, James Pustejovsky, and Robert Gaizauskas. New York: Oxford University Press, pp. 559–573.

Lin, Jo-Wang (2003). "Temporal reference in Mandarin Chinese". In: *Journal of East Asian Linguistics* 12, pp. 259–311. URL: https://doi.org/10.1023/A:1023665301095.

Mani, Inderjeet (2005). "Chronoscopes: a theory of underspecified temporal representations". In: *Schloss Dagstuhl Seminar. 05151 - and Reasoning about Time and Events.* URL: https://doi.org/10.1007/978-3-540-75989-8%5C_9.

Mani, Inderjeet, Ben Wellner, Marc Verhagen, and James Pustejovsky (2007). *Three approaches to learning TLINKs in TimeML.* Tech. rep. CS-07-268. Waltham, MA: Computer Science Department, Brandeis University.

Mani, Inderjeet, Ben Wellner, Marc Verhagen, Chong M. Lee, and James Pustejovsky (2006). "Machine learning of temporal relations". In: *Proceedings of the 44th Annual Meeting of the Association for Computational Linguistics (COLING-ACL).* Sydney, Australia, pp. 753–760. URL: https://doi.org/10.3115/1220175.1220270.

Mani, Inderjeet (2010). *The Imagined Moment.* Lincoln, NE: University of Nebraska Press.

Mani, Inderjeet and James Pustejovsky (2012). *Interpreting Motion: Grounded Representations for Spatial Language*. New York: Oxford University Press.

Mani, Inderjeet, James Pustejovsky, and Robert Gaizauskas (2005). *The Language of Time: A Reader*. New York: Oxford University Press.

Meister, Jan Christoph (2005). "Tagging time in Prolog: the temporality effect project". In: *Literary and Linguistic Computing* 20, pp. 107–124. URL: https://doi.org/10.1093/llc/fqi025.

Mostafazadeh, Nasrin, Nathanael Chambers, et al. (2016). "A corpus and evaluation framework for deeper understanding of commonsense stories". In: *Proceedings of the North American Chapter of the Association for Computational Linguistics: Human Language Technologies (NAACL-HLT'2016)*, pp. 839–849. URL: https://aclanthology.org/N16-1098.pdf.

Nebel, Bernhard and Hans-Jurgen Burckert (1995). "Reasoning about temporal relations: a maximal tractable subclass of Allen's interval algebra". In: *Journal of the Association for Computing Machinery* 42.1, pp. 43–66. URL: https://doi.org/10.1145/200836.200848.

Nieto, Marta Guerrero, Roser Sauri, and Miguel A. Bernabe (2011). "ModeS TimeBank: a modern Spanish TimeBank corpus". In: *Procesamiento del number 47 septiembre de 2011*. Lenguaje Natural, Revista, pp. 259–267.

Ning, Qiang, Hao Wu, and Dan Roth (2018). "A Multi-Axis Annotation Scheme for Event Temporal Relations". In: *Proceedings of the 56th Annual Meeting of the Association for Computational Linguistics (ACL'2018)*. URL: https://aclanthology.org/P18-1122/.

Nordlinger, Rachel and Louisa Sadler (2004). "Nominal tense in a crosslinguistic perspective." In: *Language* 80.4, pp. 776–806. URL: https://doi.org/10.1353/lan.2004.0219.

Pan, Feng, Rutu Mulkar-Mehta, and Jerry R. Hobbs (2011). "Annotating and learning event durations in text". In: *Computational Linguistics* 37.4, pp. 727–752. URL: https://doi.org/10.1162/COLI%5C_a%5C_00075.

Pustejovsky, James, Patrick Hanks, et al. (2003). "The TIMEBANK corpus". In: *Proceedings of Corpus Linguistics 2003*. Lancaster, UK, pp. 647–656.

Pustejovsky, James, Bob Ingria, et al. (2005). "The specification language TimeML". In: *The Language of Time: A Reader*. Ed. by Inderjeet Mani, James Pustejovsky, and Robert Gaizauskas. New York: Oxford University Press, pp. 549–562.

Quinn, Arthur (1982). *Figures of Speech*. Layton, Utah: Gibbs Smith.

Reichenbach, Hans (1947). "The tenses of verbs". In: *Elements of Symbolic Logic*. New York: MacMillan, pp. 287–298.

Ryan, Marie-Laure (1991). *Possible Worlds, Artificial Intelligence and Narrative Theory*. Bloomington: Indiana University Press.

Sauri, Roser and James Pustejovsky (2012). "Are you sure that this happened? Assessing the factuality degree of events in text". In: *Computational Linguistics* 38.2, pp. 1–39. URL: https://doi.org/10.1162/COLI%5C_a%5C_00096.

Sauri, Roser and Toni Badia (2012). *Catalan TimeBank 1.0. The Linguistic Data Consortium Corpus Catalog. LDC2012T10*. URL: http://www.ldc.upenn.edu/Catalog/catalogEntry.jsp?catalogId=LDC2012T10.

Sauri, Roser, Robert Knippen, Marc Verhagen, and James Pustejovsky (2005). "Evita: a robust event recognizer for QA systems". In: *Proceedings of the Human Language Technology Conference (HLT'05)*. Vancouver, Canada. URL: https://doi.org/10.3115/1220575.1220663.

Spreyer, Kathrin and Anette Frank (2008). "Projection-based acquisition of a temporal labeller". In: *Proceedings of the Third International Joint Conference on Natural Language Processing (IJCNLP'2008)*. Hyderabad, India, pp. 489–496. URL: https://aclanthology.org/I08-1064/.

Underwood, Ted (2018). "Why Literary Time is Measured in Minutes". In: *English Literary History* 85.2, pp. 341–365. URL: https://www.jstor.org/stable/26797631.

Underwood, Ted (2023). *Using GPT-4 to measure the passage of time in fiction*. Personal Blog. URL: https://tedunderwood.com/2023/03/19/using-gpt-4-to-measure-the-passage-of-time-in-fiction/.

Verhagen, Marc, Roser Sauri, Tommaso Caselli, and James Pustejovsky (2010). "SemEval-2010 Task 13: TempEval-2". In: *Proceedings of the 5th International Workshop on Semantic Evaluation (SemEval-2)*. Uppsala, Sweden, pp. 57–62. URL: https://aclanthology.org/S10-1010/.

Vilain, Marc, Henry Kautz, and Peter van Beek (1989). "Constraint propagation algorithms for temporal reasoning: a revised report". In: *Readings in Qualitative Reasoning about Physical Systems*. Ed. by Daniel S. Weld and Johan de Kleer. San Francisco: Morgan Kaufmann, pp. 373–381.

Webber, Bonnie L. (1988). "Tense as discourse anaphor". In: *Computational Linguistics* 14.2, pp. 61–73.

Yauney, Gregory, Ted Underwood, and David Mimno (2021). "Computational prediction of elapsed narrative time". In: *Proceedings of the 2021 Conference on Empirical Methods in Natural Language Processing,* URL: https://aclanthology.org/2021.emnlp-main.26/.

Yoshikawa, Katsumasa, Sebastian Riedel, Masayuki Asahara, and Yuji Matsumoto (2009). "Jointly identifying temporal relations with Markov logic". In: *Proceedings of the 47th Annual Meeting of the Association for Computational Linguistics (ACL'2009), Singapore*, pp. 405–413. URL: https://aclanthology.org/P09-1046/.

Zhang, Shuai, Lijie Wang, Xinyan Xiao, and Hua Wu (2022). "Syntax-guided Contrastive Learning for Pre-trained Language Model". In: *Findings of the Association for Computational Linguistics (ACL'2022)*, pp. 2430–2440. URL: https://aclanthology.org/2022.findings-acl.191/.

Zwaan, R.A. (1996). "Processing narrative time shifts". In: *Journal of Experimental Psychology: Learning, Memory and Cognition* 22, pp. 1196–1207.

Story Generation

6

A story has no beginning or end: arbitrarily one chooses that moment of experience from which to look back or from which to look ahead.

—Graham Greene

6.1 Introduction

As mentioned in Chap. 1, narrative generation has been decomposed into at least two stages, producing a plot, and then generating text from the plot. The fabula is often created as a byproduct of planning the plot, and is instantiated in terms of the events of the entire narrative in chronological and causal order prior to any verbalization thereof. Story generation, unlike story analysis, has its own set of structuring and, as we shall see, substantial background knowledge requirements. The approaches to story generation can be fully automatic with zero-shot or few-shot prompting, or they can involve co-authoring, via varying degrees of prompt engineering by users. In either case, most of the stories that have been studied for research purposes have tended to be short, like those in the ROC corpus. While long-form stories running from a few thousand words to novellas or novels have been generated by automatic systems, they are not likely to be of good quality, given some of the issues that arise even with generation of micro-fictions. Nevertheless, many of the narrative structures and character evolutions we expect from reading a story are hard to arrive at when dealing with five-sentence stories, but practical as well as scientific considerations suggest we are better off getting the short ones right before trying the longer ones.

In this chapter, we will first look at planning mechanisms that guide the formulation of a story. Historical story generation approaches focused on characters and their goals, creating plans relying on hand-crafted commonsense knowledge to achieve those goals, and generating texts from it. Other approaches used case-based reasoning, leveraging pre-existing story fragments and then generalizing them as needed. Included among these are systems that made use of narrative functions. To control the narrative better, systems started representing narrative goals in addition to character goals. I also describe interactive narrative, where the audience or players can shape the narrative, and may take on the role of a character, and where models of the audience can be used to control outcomes. Next, I examine temporal generation, which leverages many of the notions related to time that we discussed at length in Chap. 5. Following that, I arrive at contemporary neural approaches to story generation. This is a rather active field, and I focus on key trends. Here I examine both the use of user-generated plots as prompts, as well as system-generated plots. An important trend is the use of repositories of commonsense knowledge, which allows one to capture some aspects of character goals and causality that were the hallmark of historical approaches. I examine the challenges and approaches to evaluation, before providing an updated NarrativeML for story generation. I also assess the status of story generation systems today. Finally, I recommend treating the code that produced the story output as a first-class object of literary analysis, in keeping with the approach of the field of Critical Code studies.

6.1.1 Planning

6.1.1.1 Introduction

In modeling the generation of stories, researchers have striven for computational mechanisms that are generic, and that have sufficient expressive power. Story generation using the scripts of Schank and Abelson (1977) discussed in Chap. 3 proved to be too brittle, and not general enough, with too much of a focus on events rather than characters. Instead, planning proved to be a more generic computational mechanism. In the planning view, all characters have goals, and plans to achieve them. Every event (or state) in the narrative is viewed as being related to a character's plan (or even a narrator's plan). Story interpretation is viewed as plan recognition, which involves non-deductive reasoning, going from observed events to conclusions, which are the top-level goals of characters or narrators.[1]

> **Are plots and plans used in human narrative generation?**
> While plot can serve as a suitable commonsense explanation of a narrative that has already been generated, it is less clear that human writers proceed by constructing a plot and then fleshing it out in sentences. Some writers (myself included) start with an initial scene that introduces the setting, with a character springing forth from that scene

---

[1] Plan recognition has been formalized as *minimum covering entailment* by Kautz (1987), where the conclusion occurs in all covering models in which the observations are true. It requires encoding large amounts of world knowledge even for trivial stories. With respect to narrative, it has been subsumed by commonsense question-answering, at which LLMs have been somewhat successful.

6.1 Introduction

> and giving rise to events and subsequent scenes. The events relate in a more organic way to the characters' attributes than would be suggested by goal-oriented theories of plot, such as Plot Units. Nor do writers start with a list of topics that cover the story to be generated; this might be true of an essay, but is not a natural fit for fiction.

The generation of a narrative fabula can be viewed as plan synthesis, or planning, which involves finding a sequence of actions that maps from the initial state to the goal state, as defined by Riedl and Young (2010). The actions have preconditions and postconditions. Let us say a character called John is broke and hungry, and that the world he inhabits has a store with an edible burger. Assume John is armed, and that he has a cat called Mia, and that there is in addition a bank conveniently at hand, in possession of money. Eating satisfies an agent's hunger, but requires that the agent be hungry and possess something edible. Buying requires that the buyer have money and the seller have something, and the result is a swap. Robbing requires that the robber have a gun, and that the institution to be robbed have money; the result is that the robber gets the money. This representation of the world will have to be used by the planner to produce a fabula. The computation involves searching for an appropriate sequence of actions for getting from the initial state to the goal state, given the many facts and rules in the world.

Clearly, John being armed and the bank having money satisfies the condition for robbing the bank, as a result of which John will have money. Then the conditions for buying the burger will be satisfied, as a result of which he will buy it, which in turn will allow him to eat it. This is an example of planning using bottom-up reasoning. Alternatively, one can proceed top-down. John wants to satisfy his hunger. The action of eating can do that, provided the agent is hungry and there is something edible available. The latter can spawn a new goal, to buy something edible, for which he needs money. To get that, he can rob a bank.

If we proceed top down, John's initial state has him hungry, with a gun, and the bank having money, and John not having it, and the store having a burger, which is edible. His goal is not to be hungry. Commonsense knowledge specifies that to eat, the preconditions are being hungry, that the thing to eat is edible, and the postcondition is not being hungry. Commonsense also specifies that to buy, the preconditions are having money, and that there is a store which has the thing one wants to buy, and the postcondition is that one has the thing, and the store has the money. Finally, to rob, the precondition is that one has a gun, and the place to rob has money, and the postcondition is that one has the money. Given the goal and these three commonsense rules, John formulates a plan to rob the bank and then buy the burger with the stolen money, and then eat the burger, satisfying his hunger.

All this lines up neatly. In fact, a bit too neatly! The particular rules of relevant commonsense knowledge conveniently happen to be at hand. They are also stated in such a form that they can easily match his goals. That may be fine for a pedagogical example, but the question arises as to where such knowledge in the right form is to come from. Attempts to encode such knowledge by hand and provide them to systems have worked only in small domains where the knowledge can be specified in this way. That hand-crafted approach falls apart when it comes to building a general-purpose, open-domain system. Commonsense, as Voltaire observed in another context, is not so common. This is a serious lacuna given that story generation seems to require, for the causal links between events to make sense, substantial dollops of commonsense knowledge. However, modern neural approaches seem to have acquired some of these dollops implicitly via their training data. We will see this implicitly acquired knowledge exposed to some extent in the pre- and post-conditions for events in auto-generated NarrativeML for stories.

We will discuss more robust sources of commonsense knowledge for story generation in Sect. 6.4.1.3. Meanwhile, here are the details of John's plan, for those who want to read the logic directly.

6.1.1.2 Planning Details*

The planning problem and its solution plan can be formalized as shown in Example 6.1. It can be seen that the plan incorporates both the character's mental states and goals as well as actions.

Example 6.1

```
Initial state:
hungry(John), bank(TheBank), store(TheStore), has(John, y), gun(y),
has(John, Mia), cat(Mia), has(TheBank, z), money(z),
not(has(John, z)), has(TheStore, The99centBurger),
edible(The99centBurger)
Goal state:
not(hungry(John))
Domain theory:
eat(x, y): pre: hungry(x), has(z, y), edible(y);
post: not(hungry(x))
buy(x, y): pre: money(z), has(x, z), has(p, y), store(p);
post: has(x, y), has(p, z)
rob(x, y): pre: has(x, z), gun(z), has(y, p), money(p);
post: has(x, p)
Plan:
rob(John, TheBank); buy(John, The99centBurger);
eat(John, The99centBurger)
```

Planning can be viewed as finding a proof for a goal given a set of initial conditions and actions. It has been formalized as propositional satisfiability (PSAT) (or constraint satisfaction), i.e., finding a model for a set of axioms, as shown by Kautz and Selman (1992). There are a variety of off-the-shelf solvers for PSAT problems.

How does this plan lead to a story? We turn to this next.

6.2 Historical Story Generation Approaches

Story generation has a long history in NLP and AI. For reasons of space, I include only a few highlights that are relevant to current problems. For an excellent compendium of historical systems output and interactions, see Bertram and Montfort (2024).

6.2.1 Character Goals

To generate a fabula as a plan, one must create characters with an initial state and a goal state. The planner uses a built-in domain theory to find a plan to get a sequence of actions to get to the goal state. The initial state, goals, and the trace of the actions carried out in the plan constitute the story fabula. Obviously, any narrative generation system has to be precise about what goes into the fabula. That is in itself an important contribution to narratology, given that narratological studies have typically not discussed fabulae at the level of the detailed propositions involved.

One of the classic story planning systems is TALE-SPIN, developed by Meehan (1976). To start the system off, the characters and geographical locations in the story world are initialized, and each character is given a goal to solve. Each goal spawns further subgoals involving specific actions. As the actions are carried out, the state of the world is updated. The story generated is essentially a trace of the goals and actions of the characters. An example TALE-SPIN story (with my elisions indicated by ...) is shown in Example 6.2.

Example 6.2
George was very thirsty. George wanted to get near some water. George walked from his patch of ground across the meadow through the valley to a river bank. George fell into the water. ... Wilma wanted George to get near the meadow. Wilma wanted to get near George. Wilma grabbed George with her claw. Wilma took George from the river through the valley to the meadow.

TALE-SPIN can include both cooperative as well as adversarial activities between characters. As pointed out by Wardrip-Fruin (2009), there is also a degree of narrative embedding. Consider the case where a third character Arthur asks George where to find honey. If George's initial state is such that he is allowed to lie, he invents a character called Ivan Bee who has honey in a beehive in a redwood tree. If Arthur trusts George somewhat, Arthur's beliefs are updated with this fictitious information. Arthur is no Scheherazade, but he has the ability to spin tales within tales.

The approach of TALE-SPIN allows for all the characters' actions to be traceable to (explicit statements of) their motives (i.e., their goals). Nevertheless, it has numerous narratological shortcomings:

- As a trace of character's goals and the actions used to solve them, TALE-SPIN's stories seem rather mechanical and lacking in many of the narrative devices we examined in Chaps. 3 and 4. They resemble running commentaries, blow-by-blow accounts of what the character thought and did.
- The ordering of events, mental or other, is always a chronology, in the terminology of Genette (1980).
- There is no variation in narrative distance or perspective. Nor is there much stylistic variation, even though TALE-SPIN could generate sentences that were syntactically complex, as in the above example.
- TALE-SPIN is also notorious for getting stuck in dead-ends, its so-called mis-spun tales, as in Example 6.3.

Example 6.3
Henry Ant was thirsty. He walked over to the river bank where his good friend Bill Bird was sitting. Henry slipped and fell in the river. He was unable to call for help. He drowned.

6.2.2 Case-Based Reasoning

The idea of leveraging data for planning narratives attempted to generalize through adaptation of previous examples. **Case-Based Reasoning** (CBR), which involves retrieving and generalizing fragments of narratives that have been stored in memory, is another generic computational mechanism (for an introduction, see Kolodner (1993)). CBR involves comparison with stored memories of previously similar situations to solve a problem, and has been used effectively for legal and analogical reasoning. Systems to understand and especially, generate narrative have thus availed of CBR, combining it with planning formalisms.

The idea of stories having their origins in previous fragments that are modified anew is not just germane to AI systems, but forms the basis for the improvisation found in human storytelling, as shown in studies of oral epics by Lord (1960) and Finnegan (1977), as well as fan fiction. The nature of the retrieval and generalization as used in CBR can be of considerable interest to corpus-based studies, for example, in finding characters across narratives who experience common predicaments that are addressed in different ways.

PROTOPROPP, developed by Peinado and Gervas (2006), generates stories using Case-Based Reasoning and Propp's Narrative Functions. The cases are indexed by the 31 Proppian functions (structured as an event hierarchy) and character types. The reader may want to consult Table 4.1 in Chap. 4 for the list of narrative functions.

PROTOPROPP generates new story fabulas from a query by retrieving similar fabulas from the case base, while randomly deleting some narrative functions not in the query (along with dependents), and adding others and dependents at random. It uses backtracking when constraints related to the participation of characters in Proppian functions are violated.

6.2 Historical Story Generation Approaches

Here, in Example 6.4, is a rather simplistic story that gets generated based on automatic modifications to *The Magic Swan Geese* fabula.

Example 6.4
Once upon a time there was a princess. The princess said not to go outside. The princess went outside. The princess heard about the lioness. The lioness scared the princess. The lioness kidnapped the princess. The knight departed. The knight and the lioness fought. The knight won the fight. The knight solved the problem of the princess. The knight returned. A big treasure to the knight.

The PROTOPROPP ontology has been linked to a Proppian annotation of folk tales by Scheidel and Declerck (2010) to support information extraction. And Grasbon and Braun (2001) describe a partially implemented interactive storytelling system based on Proppian functions.

Despite the issues we encountered in Chap. 4 with annotation of narrative functions, Propp's morphology, being formal and precise, is still rather attractive for generation, at least of folk tales and simple narratives. Propp himself (Propp 1968, pp. 111–112) suggested presciently how his morphologies could be used for generation:

In order to create a tale artificially, one may take any A, then one of the possible B's then a C ↑, followed by absolutely any D, then an E, the one of the possible F's, then any G, and so on. In doing this, any elements may be dropped, or repeated three times, or repeated in various forms. If one then distributes functions according to the dramatis personae of the tale's supply or by following one's own taste, these schemes come alive and become tales. Of course, one must also keep motivations, connections, and other auxiliary elements in mind.

6.2.3 Narrative Goals

Planned stories need a better way to control narrative outcomes, rather than leaving the story to unravel based on the successes and failure of characters' plans. Further, there are often authorial intrusions (not only those of the extreme *Tristram Shandy* variety) that cannot be modeled in terms of character goals.

UNIVERSE, developed by Lebowitz (1985), is a system that uses narrative goals as well as character-driven goals to plan narratives. It works in the domain of soap opera, specifically *Days of Our Lives*. The system is initialized using a character creation cycle, where each character wins and loses spouses, has children, and possibly dies. Each character is given a set of traits, based on stock characters. If UNIVERSE is given a narrative goal that calls for a couple to divorce, it will insert relationship obstacles into the narrative. A typical goal involves *churning* couples, keeping them separated by new obstacles each time the previous set is cleared up.

UNIVERSE makes use of CBR. An existing plot fragment, represented as a plan, may be used or generalized to achieve this. These fragments include typical constructs relevant to such soap-operas, such as LOVERS-FIGHT, STEAL-CHILD, COLLEAGUE-AFFAIR, JOB-PROBLEM, etc. Example 6.5, from Wardrip-Fruin (2009), shows the goals of churning Liz and Neil and getting Neil and Renee together, and the resulting narrative trace (the fabula produced is prefixed by =>). As the narrative unfolds, Liz will be threatened, Renee will seduce Neil, and Liz will get divorced.

Example 6.5

```
(tell (((churn liz neil)(together renee neil))))

working on goal (CHURN LIZ NEIL)

Several plans to choose from
FORCED-MARRIAGE LOVERS-FIGHT JOB-PROBLEM
using plan FORCED-MARRIAGE
working on goal (DO-THREATEN STEPHANO LIZ ''forget it'')
using plan THREATEN
=> STEPHANO threatens LIZ: ''forget it''

working on goal (WORRY-ABOUT-NEIL)
using plan BE-CONCERNED
Possible candidates MARLENA JULIE DOUG ROMAN
      DON CHRIS KAYLA
Using MARLENA for WORRIER
=> MARLENA is worried about NEIL

working on goal (TOGETHER * NEIL)
Several plans to choose from SEDUCTION DRUNKEN-SNEAK-IN
SYMPATHETIC-UNION JOB-TOGETHER
using plan SEDUCTION
Possible candidates DAPHNE RENEE
Using RENEE for SEDUCER
=> RENEE seduces NEIL

working on goal (ELIMINATE STEPHANO)
Several plans to choose from ATTEMPTED-MURDER EXPOSE
using plan ATTEMPTED-MURDER
Using ALEX for KILLER
=> ALEX tries to kill STEPHANO

working on goal (DO-DIVORCE TONY LIZ)
using plan DIVORCE
=> LIZ and TONY got divorced

working on goal (TOGETHER LIZ NEIL)
no acceptable plans
```

UNIVERSE is creative but impractical:

- It was never fully completed, and it does not do surface text generation.
- It is tied to the domain of soap opera, and severely limited by its knowledge of previous cases in memory.
- More importantly, it is far too author-driven: no actions can occur unless they satisfy a narrative goal.

Like UNIVERSE, the system MINSTREL of Turner (1993) uses CBR to tell stories. Given a particular goal, it creates a story by selecting and modifying a previous plan (schema) from memory. It is tied to the domain of Arthurian legend. As an example, if the system is given the goal for a knight to kill himself, it may try to find a plan where the knight can do so by drinking a potion (Perez et al. 2004, Wardrip-Fruin 2009). Let us say it finds only one somewhat similar plan, where a princess drinks a potion to make herself ill. By reasoning that being killed is similar to being injured, and that injury can result in death, the goal will be altered to having the knight injure himself. As there is still no plan matching this goal, MINSTREL will try to generalize the goal to find anyone doing something for self-injury, rather than just a knight. This time, the princess schema is matched and MINSTREL chooses a scene where the knight kills himself by drinking a potion.

As with UNIVERSE, MINSTREL is limited by its knowledge of previous cases in memory, but it is able to generalize to partially overcome that limitation. It has no idea of what sort of outcomes are satisfactory; for example, a knight can adapt a plan from a dragon, killing and eating a princess.

Collectively, classic symbolic systems such as TALE-SPIN, UNIVERSE, and MINSTREL all suffer from well-known problems:

- They are brittle, with careful crafting required for each story domain.
- Knowledge acquisition for these and other systems remains a challenge.
- They also lack any model of audience satisfaction, and thus many automatically generated action sequences and narrative outcomes may be undesirable. To address this recent developments related to narrative planning include the modeling of intentionality of actions to generate more plausible fabulae by Riedl and Young (2010).

For more on these systems, see Gervas (2014).

6.2.4 Planning for Interactive Narrative

As mentioned in Chap. 3, the **audience** or **actual reader** is crucial to storytelling. The audience may be passive, as with spectators witnessing a performance or readers perusing a book; or it may be an active participant in the storytelling, as in the case of traditional call-and-response oral narratives or in interactive computational storytelling environments.

In the latter case, the audience or player can shape the course and outcome of the narrative, and may take on the role of a character. An overall plan for the narrative thus has to take into account the role of such an audience. This is one of the factors that makes games with interactive narratives rather different from traditional narratives. While Murray (1997) has argued that video games, as 'cyberdrama', fall squarely within the scope of storytelling, *ludologists* such as Aarseth (1997) have emphasized such differences between games and traditional narratives, while acknowledging that there is substantial overlap between the two. It is not my purpose here to survey this debate, but two key problems are relevant from the point of view of the plans of intentional agents:

1. *Replanning*: How can the system dynamically replan the narrative given particular audience moves? Here player interactions can cause earlier plans to be revised.
2. *Outcomes*: How can the system steer an interactive narrative toward satisfactory outcomes for the audience? This is required since a player, even if he doesn't run amok, can lead a narrative toward dead-ends or unsatisfactory outcomes

Regarding replanning, I will focus on the use of **reactive planning** for interactive narrative and drama in the FAÇADE system of Mateas and Stern (2005). The story world has a room with two characters, Trip and Grace (who are modeled on George and Martha in Edward Albee's *Who's Afraid of Virginia Woolf*), along with the player, who can be outside or in the room. FAÇADE relies on the reactive planning language ABL (A Behavior Language) of Mateas and Stern (2002), which is in turn derived from the *believable agent* planning language HAP of Loyall and Bates (1991).

FAÇADE's Drama Manager differentiates dramatic actions from other events in terms of the use of *beats*, where a *beat* is the smallest unit of dramatic action (not to be confused with the short pause in a dialogue). In the screenwriting prescriptions of McKee (1997), beats are used to assemble scenes. Each scene involves events that change *values* from positive to negative or vice-versa, where *value* is a significant change in certain properties of individuals and their relationships. For example, (ibid., p. 34), in a time of terrible drought, rainfall would mark a change of value (from death to life) in people's lives. In FAÇADE, each beat is chosen based on its preconditions, the player's moment-by-moment interaction, the beats so far, and an overall Freytagian dramatic arc. An indirect narratological contribution of Mateas and Stern (2005) is thus to highlight the importance of beat-based sequencing to the overall narrative arc.

Although the scheme of McKee (1997) is highly informal and not entirely consistent, his account of narrative structure in films involves a series of two to five scenes forming a sequence. Sequences in turn form acts, which culminate in a climactic scene, with a story being made up of acts. The critical assessment of such screenwriting concepts for their narratological import remains to be carried out, but their relevance is apparent in FAÇADE.

FAÇADE is of narratological interest because it shows how narratives can be replanned dynamically as the audience responds. These real-time constraints have been largely ignored

6.2 Historical Story Generation Approaches

in narratological studies of improvisational theater. It nevertheless has a couple of salient shortcomings:

- As will be seen, a large number of specific plans have to be written to synchronize the interactions and the dialogue between characters given the presence of the player.
- Further, this sort of detailed reactive planning involves heavy scripting of the narrative that mitigates against the ability to plan more open-ended narratives.

The next section goes into computational details of FAÇADE, and may be skipped by those uninterested in such classic replanning formalisms.

6.2.4.1 Replanning Details*

FAÇADE's planning architecture consists of plan memory (all plans known to an agent) stored as production rules involving goals and their preconditions.[2] The active plan tree (involving goals active for an agent) is adjusted at runtime to remove goals that have been spontaneously achieved (this is implemented via success tests) and to remove active plans that are no longer applicable (implemented via context condition tests). This adjustment can cause a failed plan to be replaced by a new plan, if any.[3]

As an example, from Mateas and Stern (2002), there are two behaviors specified for a system agent to satisfy the goal of opening a door: first, yelling for the guest to come in and waiting for them to open the door, and second, walking to the door and opening it. If both behaviors are applicable given the preconditions, the more specific one is chosen. This is shown in Example 6.6.

Example 6.6

```
sequential behavior OpenDoor() {
    precondition {
      (Knock doorID :: door)
        (Pos spriteID == door pos :: doorPos)
        (Pos spriteID == me pos :: myPos)
          (Util.computeDistance(doorPos, myPos) > 100)
    }
      specificity 2;
// Too far to walk, yell for guest to come in
      subgoal YellAndWaitForGuestToEnter(doorID);
}
```

---

[2] The left-hand side of a rule is a goal and precondition, while the right-hand side has the rule context, specificity, and plan expression. The plan expression involves parallel or sequential arrangement of steps, where each step contains a goal expression, the goal priority, and the goal success test.

[3] Choices between multiple plans are arbitrated using goal priority. Plans can also be dynamically extended, implemented by goal sensitivity.

```
sequential behavior OpenDoor() {
 precondition {
    (Knock doorID :: door)
    }
    specificity 1;
    // Default behavior - walk to door and open
    ...
}
```

Regarding its use of beats, consider, in Example 6.7, three beat goals (from Mateas and Stern 2002) of (i) first opening the door and greeting the player, (ii) then yelling for Grace, and (iii) finally inviting the player into the apartment:

Example 6.7

```
sequential behavior BeatGoals() {
    with (persistent when\_fails)
        bgOpenDoorAndGreetPlayer();
    with (persistent when\_fails)
        bgYellForGrace();
    with (persistent when\_fails)
        bgInviteIntoApt();
}
```

The beat goals are monitored by beat handlers. When the player refers to Grace, the beat handler for Trip ignores such references if Trip is in the middle of beat goal (ii) of yelling for Grace. Otherwise, Trip interrupts himself with "Oh, yeah," performs the beat goal (ii) within the handler, or, if that goal has already happened, says "She's coming. I don't know where she's hiding." The beat handlers are such that player interaction can cause beat goals to be interrupted, reordered, and responded to in a way dependent on what has happened in the beat so far.

6.2.4.2 Generating Preferred Outcomes

As mentioned in Chap. 3, the audience has strong beliefs about the narrator and characters in the novel. In narrative understanding, the audience *is* the interpreting system. However, in generating narrative, interactive or otherwise, a **user model** (or reader or audience model) is desirable. The information in such a model needs to include how the user reacts to outcomes of particular events.

How do players or users of intelligent narrative systems respond emotionally to narratives? Models of possible user reactions can be used in an intelligent narrative system in an evaluation function to prefer choices in the narrative that will have a maximum impact at particular times on the user. This could be especially useful in pruning possible paths in a narrative generator. It can be more or less critical in interactive narrative systems, e.g., where the provision of excessive autonomy to the player can weaken authorial control and intent. Such an evaluation function could restrict user choices to those that are predicted to have particular types of impact on the user.

Unfortunately, reasoning about preferred outcomes has not been a strong suit in planning formalisms. There has, however, been some related work in interactive narrative generation. Cheong (2007) generates stories judged to be suspenseful by modeling the reader's reasoning about limitations and conflicts involving a protagonist's goals, based on psychological insights from Gerrig and Bernardo (1994) and earlier work by Cheong and Young (2006). Cheong (2007) describes a narrative generator that uses a reader model to create potentially more suspenseful stories. She manipulates two specific suspense parameters:

1. a level-of-suspense heuristic where the suspense increases as the number of planned solutions in the reader model for a particular protagonist's goal decreases (reflecting the finding from Gerrig and Bernardo (1994) that subjects report increased suspense as the number of plans available to a protagonist dwindles);
2. an (action-specific) suspense heuristic where the potential suspense for an action increases with the number of perceived obstacles (measured by the difference in the reader model between the number of effects that thwart a protagonist's goal and the number that support it).

Her system evaluation shows that these manipulations do alter the reader's suspense level.

6.3 Temporal Generation

Let us briefly consider time in natural language generation, where given the fabula with its ordering, the discourse order must be computed. This is in fact a simpler problem than in understanding, when one has to go from the discourse to the fabula. The historical system STORYBOOK of Callaway (2000) takes as input a set of propositions (who did what to whom in the Red Riding Hood domain), including characters, scenes, events, and a partial ordering of events. It produces a fabula, plans sentences for the discourse, fine-tunes them, and realizes them with a sentence generator using an English lexicon and grammar. The system does not plan the fabula but uses a graph-based programming formalism called a finite-state machine to select a single path through predefined event and scene sequences. It ensures consistency, pruning inconsistent paths and using safe scenes. In converting fabula to discourse, STORYBOOK allows focus on specific characters and events, specifying roles (hero, villain) and adjusting narrator perspective and voice (first, second, third person). Defaults and user specifications guide parameter values.

For Little Red Riding Hood, the generated timeline involves the girl meeting the wolf, gathering wildflowers, and the wolf rushing ahead in parallel, using an interval calculus representation. STORYBOOK decides the tense and when to shift it, especially in dialogue where tense may differ from the narrative. It plans and realizes sentences with fluent tense shifts using a grammar and lexicon of English.

Example 6.8
Little Red Riding Hood had not gone far when she met a wolf. "Hello", greeted the wolf, who was a cunning-looking creature. "Where are you going?" "I am going to my grandmother's house", Little Red Riding Hood replied.

However, STORYBOOK relies far too heavily on built-in defaults and user specifications for its system parameters. See Lonneker (2005) for a detailed critique.

In contrast to the use of grammar-driven sentence realization, many historical systems that generated dialogue and short text snippets, used template-driven text generation. Here, the templates map non-linguistic input directly to the linguistic output form, sacrificing linguistic generalization for rapid prototyping. Perhaps the most time-cognizant of all the generation systems is the interactive fiction system CURVESHIP of Montfort (2011). It varies the narrator's speech time and the reference time with respect to the event time, to decide on the particular tense. It also makes use of temporal adverbials and conjunctions like 'then" and 'before" to express temporal relations, as shown in Example 6.9.

Example 6.9
Your senses were humming as you viewed the broad, circular, encircling Plaza of the Americas. The morning had concluded. It was midday then.

CURVESHIP implements all of Genette's seven orderings, using the following heuristics:

1. Achrony: order events at random.
2. Analepsis: select the most salient event from the first time the focalizer encountered this character, or else select the most salient events that the focalizer has seen happen in this room in the past, up to three of them.
3. Chronicle: sort events in chronological order.
4. Prolepsis: pick 'inevitable' events such as the sun going down, or nuclear missiles arriving.
5. Retrograde: sort events in reverse chronological order.
6. Syllepsis: use whenever the adventurer is entering a new area, or defeating a monster, or acquiring a treasure.
7. Zigzag: all the events in a single physical location are narrated in the "now," and then the corresponding events in the "then," and so on for each location.

What CURVESHIP does not do, however, is decide when it is most appropriate to use one type of ordering rather than another. As such, discourse-ordering choices such as these can benefit from corpus-based narratological studies.

6.4   Contemporary Narrative Generation Trends   199

GPT-4o has little difficulty in generating different narrative orderings, e.g., *Zigzag* for our little ROC (Mostafazadeh et al. 2016) spaghetti story in Example 6.10.

Example 6.10

> **User Prompt**: Zigzag is a narratological term for where the narrator alternates between two times, e.g., 'now" and once," as in:
> *Sometimes passing in front of the hotel he remembered the rainy days when he used to bring his nursemaid that far, on a pilgrimage. But he remembered them without the melancholy that he then thought he would surely some day savor on feeling that he no longer loved her..*
> Here is a story that I would like you to convert to zigzag narrative ordering.
> *Tina made spaghetti for her boyfriend. It took a lot of work, but she was very proud. Her boyfriend ate the whole plate and said it was good. Tina tried it herself, and realized it was disgusting. She was touched that he pretended it was good to spare her feelings.*
> **Machine**: *As Tina watched her boyfriend finish the last bite of spaghetti, she thought back to the hours she'd spent preparing it, carefully chopping ingredients and stirring the sauce. He smiled, telling her it was good, and she felt a surge of pride, remembering how the aroma had filled the kitchen, making her eager to share the meal. Curious, she took a bite herself and was hit with the realization that it was terrible—over-seasoned, the texture all wrong. But earlier, when she placed the plate in front of him, she had been so hopeful, watching as he took his first bite with anticipation. Now, she saw how he had eaten it all without complaint, and the realization that he pretended to enjoy it just to spare her feelings filled her with quiet gratitude.*

These Foundation Models too aren't as yet skilled enough to know when to choose such orderings.

6.4   Contemporary Narrative Generation Trends

Contemporary research has moved away from the domain-specific and template-driven methods that characterized historical approaches, relying in large part on the neural machinery we have been discussing. Since sentence generation is carried out by the LLM, computational narrative approaches have focused mainly on the generation of plots and other higher-level constraints to guide the LLM to generate a coherent narrative. Of course, the LLM itself is not that easy to control at inference time, except by way of prompts and the few parameters that the developers have graciously exposed in the API. Research has attempted to inject constraints into the next token prediction (for example, by lexically constraining it with specific keywords). Since wholly retraining LLMs is extremely expensive, fine-tuning is often adopted for enhancing performance.

6.4.1 Plot Planning

6.4.1.1 User-Generated Plot Prompts

Engineering the prompts by carefully choosing words that should be included in the output, along with guidelines for length, quality, style, and so forth can also make a big difference, as can specific instructions to imitate well-known authors. However, this is largely a matter of trial-and-error. We have already mentioned the use of prompts to plan suspenseful stories in Xie and Riedl (2024). Example 6.11 shows a prompt provided to GPT-o1-preview for story plot generation, and the answer (for reasons of space, I omit the LLM's answers to each prompt):

Example 6.11

> **User Prompt 1**: I would like to generate a plot for a detective story. First, consider the setting. The story takes place in a country house. Describe the atmosphere and features of the house that will set the tone for the story.
> **User Prompt 2**: The victim is a well-known figure with a shadowy past. What details about his life will make him an interesting character?
> **User Prompt 3**: Who are the main suspects in the murder? Consider several characters with potential motives.
> **User Prompt 4**: Describe the inciting incident. What event triggers the investigation into the murder?
> **User Prompt 5**: What are the major turning points in the investigation? List a few events that alter the trajectory of the plot or reveal new clues.
> **User Prompt 6**: Think about the climax of the story, when the detective confronts the killer. How does the detective figure out who the murderer is, and how does the confrontation unfold?
> **User Prompt 7**: Finally, how does the mystery resolve? What happens after the revelation, and how does the story end?
> **Machine**: *Set in a secluded, mist-enshrouded country manor, a renowned philanthropist with a shadowy past hosts a stormy evening gathering attended by four key figures: the estranged heir resentful over disinheritance; a business partner fearing exposure of embezzlement; a reclusive neighbor with old grievances linked to a past scandal; and a loyal assistant secretly harboring resentment over exploitation.*

6.4 Contemporary Narrative Generation Trends

> *During a sudden power outage, the victim is found murdered in the library with an antique letter opener.*
> *The investigation reveals a hidden passage from the library, an altered will favoring the assistant, and forensic evidence indicating the murder occurred before the blackout, undermining the suspects' alibis.*
> *The detective deduces that the business partner, who knew the house intimately and was about to be exposed by the victim, is the murderer.*
> *In a climactic confrontation, the business partner confesses. Justice is served as they are arrested, the heir finds closure, the assistant honors the victim's legacy, and the detective departs, leaving the manor's secrets unveiled.*

This example does demonstrate that the language of plots and other narrative elements is just language (some of it with politically correct pronoun use), and for completing such linguistic sequences, autoregressive next-word prediction is not a bad fit.

To automatically generate a very short story from a COT- or human-generated plot is not that difficult. Feeding the above plot to various Foundation Models returns interesting results, albeit atrociously over-the-top at times[4] in style. Claude 3.5 Sonnet, in particular, includes dialogue, as seen in Example 6.12:

Example 6.12

> **User Prompt**: Generate a 500-word story given the above plot outline.
> **Machine**: *Mist enveloped Thornwood Manor as thunder rumbled ominously. Four figures arrived, each harboring dark secrets: James Thornwood, the estranged son; Victoria Sinclair, the nervous business partner; Harold Blackwood, the brooding neighbor; and Sarah Mills, the seemingly loyal assistant.*
> *Lord Edmund Thornwood, once a celebrated philanthropist now shrouded in mystery, greeted them in the grand foyer. As lightning flashed, he led them to the drawing room, tension palpable in the air.*
> *Suddenly, a thunderous crack plunged the manor into darkness. Amid the chaos, Sarah's trembling voice called out, "Lord Thornwood? You need to come quickly."*
> *In the library, candlelight revealed a gruesome scene: Lord Thornwood lay dead, an antique letter opener protruding from his chest. Detective Inspector Alastair Grey arrived, his keen eyes taking in every detail. As the storm raged, he began unraveling a web of motives.*
> *James, disinherited years ago, seethed with resentment but denied involvement. Victoria fidgeted nervously, her alibi weak. Harold spoke cryptically of old grievances, while*

---

[4] For traditional examples of over-the-top, AI-free writing, see https://www.bulwer-lytton.com/.

> *Sarah professed unwavering devotion to the deceased. Grey's investigation uncovered startling revelations: a hidden passage behind a bookcase, connecting the library to the wine cellar; forensic evidence indicating the murder occurred before the blackout; and a newly amended will favoring Sarah.*
>
> *The crucial clue came from dust patterns on the library floor's footprints made by expensive Italian leather shoes, worn only by Victoria Sinclair. Confronted, Victoria crumbled. "He was going to ruin everything," she confessed. "The embezzlement, the offshore accounts—he'd discovered it all." As Victoria was led away, the remaining occupants grappled with newfound truths. James realized the complexity of his father's legacy, while Sarah vowed to honor Lord Thornwood's philanthropic vision.*
>
> *Inspector Grey departed, leaving behind a manor stripped of its darkest secrets. Yet as mist continued to shroud Thornwood, he couldn't shake the feeling that some mysteries would remain forever hidden, waiting for the next unwary soul to stumble upon them.*

The generated story has some discourse-level flaws at the beginning, in the use of definite articles to introduce new characters, but it does seem to be able to cover all the plot points. Claude 3.5 Sonnet, like other tools I tried, didn't manage to hit the length target. As the story to be generated gets longer, standard detective-story devices like false clues are introduced by the LLMs. However, it doesn't take long before incoherence arises. For example, at 2000 words, GPT-4o devotes most of the story to the setting, indicial actions, and the lead-up to the murder, leaving very little room for the detective and no room for the unraveling of the whodunit, analogous to the mis-spun tales of TALE-SPIN. Clearly, the ability to accommodate target length appropriately, a problem that has been solved in automatic summarization using, e.g., *knapsack algorithms*, has not been tackled by present-day generation systems. Nor has the problem of *long-distance narrative dependencies* been addressed by these Foundation models. To address it, we need to integrate story planning with sentence generation at a more fine-grained level.

6.4.1.2 System-Generated Plot Prompts

Autogenerating Plots

Rather than generating a plot with ad hoc user prompts as in Example 6.8, research (Ye et al. 2023) has explored methods for auto-generating prompts by representing causal dependencies. For instance, prompted with the text corresponding to the story ending "Sally buys a gun", the LLM produces preconditions like "Sally is at the store" and "Sally has money". (Few-shot learning is used to break down preconditions into cases like required items, locations, and prior events.) The system generates events for each precondition, resulting in a partially-ordered graph that is flattened into a plot for the LLM. Evaluated on the ROC dataset (Mostafazadeh et al. 2016), coherence was measured by checking if story events were causally linked.

6.4 Contemporary Narrative Generation Trends

Machine-Learning of Plots

Another approach is to automatically learn plots from a dataset of stories. By clustering sentences across crowdsourced stories, the system of Li et al. (2013) detects the most frequent events (clusters of related sentences or phrases) and learns their most likely ordering constraints. These inferences are carried out within and across stories; a 30-node plot graph for a bank robbery, for example, is generated from 60 crowdsourced examples. The system generates stories by a random walk across the graph, adding events stochastically to the story such that no constraints (e.g., for precedence or mutual exclusion) are violated. The story completes when an ending event is reached. The generation algorithm can produce over a million linearizations for their bank robbery graph! The evaluation based on post-edit measures by humans shows that stories generated are comparable in quality to simple stories authored by untrained humans. More direct evaluations of output quality were not conducted.

Simplified Plot Representations

Given the difficulties in specifying or harvesting plots, recent work has explored simplified representations of plot. In these approaches, the representations are limited to totally ordered event sequences, losing the information about goals and preconditions. Using the ROC corpus as training data, the plan-and-write Yao et al. (2019) approach learns, given a short title (e.g., "Computer"), a sequence of keywords, one from each sentence of a corpus story, that they call a storyline (e.g., "needed → money → computer → bought → happy"). That storyline, which is reminiscent of a Narrative Event Chain (NEC), is converted into a prompt sequence for output story generation, producing stories rather like the ROC ones. To avoid the common problem of repetition within and across generated stories, they implement a diversity metric in their training objective. While the generated sentences are locally coherent, the overall story may not be, with logical inconsistencies remaining a problem. Instead of using word sequences for storylines, the approach of Fan et al. (2019) generates, given an initial prompt, a sequence of predicates and arguments with semantic role labels. This refined storyline, called an action plan, is then turned into sentences with placeholders for entities, which are then fleshed out with specific references.

Instead of limiting the initial prompt that describes the overall topic or title, some approaches have provided information from the end, as we have seen, or, as in research by Ippolito et al. (2019), feeding the beginnings and endings of stories for the LLM to generate the middle, which can be useful in a co-authoring setting. Other work such as Peng et al. (2018) has incorporated automatically analyzed story-ending valences (limited to happy or sad endings) from training data and used them as additional conditions to control the story generation for language models.

6.4.1.3 Integrating Commonsense Knowledge

Another key trend is the integration of commonsense knowledge via neurosymbolic approaches. The use of specific domain knowledge within such approaches is also gaining ground, especially for co-authoring environments in interactive or other applications, as in the work of Kelly et al. (2023). However, as we discussed in Sect. 6.1.1.1, encoding commonsense knowledge by hand is only possible for a small domain, and proves brittle in the large. Therefore our focus here is on the use of much broader commonsense knowledge. As we shall see, a variety of commonsense data resources are available that have been integrated with LLMs.

These knowledge sources, when in the form of knowledge graphs, can be integrated with LLMs in a variety of ways. The graphs can be used to generate separate embeddings, which are accessed when, say, a word in the input has information in the graph. The methods discussed in Chap. 2 are relevant here: Fine-tuning of the LLMs can be performed on data from the knowledge graph, whether in structured form or converted into sentences using LLMs. Likewise, a RAG approach can be used to access the knowledge graph, with the retrieved results fed into prompts, or using intermediate-layer integration, as in the case of Borgeaud et al. (2022) discussed in Chap. 2, where retrieval blocks with word embeddings from retrieved passages are interleaved with transformer blocks.

A key type of resource comes in the form of LLMs trained on large knowledge graphs (Bosselut et al. 2019). These LLMs can express their learned knowledge in terms of semantic relations. For example, when such a trained LLM is given "take a nap Causes", it could complete the pattern by generating "have energy". COMET-2020, from Hwang et al. (2021), is a Transformer trained on relationship triples from their ATOMIC-2020 KB (which has 1.33M commonsense triples across 23 commonsense relations, acquired by crowdsourcing and merging with prior KBs).

Applying this knowledge to narrative generation, the work of Peng et al. (2022) begins with a prompt sentence featuring a specific character to focus on, generating candidate continuations that are assessed against commonsense attributes inferred for that character. The top-matching continuation, which aligns with inferred attributes (e.g., expected emotions or reactions), is appended to the story, iteratively updating the prompt until the story reaches the desired length. By chaining the preconditions (facts or motivations required for the current sentence) with the post-conditions of the previous sentence (effects or new intentions generated by prior events), the system checks for coherence to ensure reader comprehension, such as inferring that person A, after receiving a million dollars from person B, would want to thank him.

6.4.1.4 Integrating Craft Knowledge

One could argue that a lot of knowledge about story generation is already being implicitly mined by LLMs because their training data includes a considerable amount of narrative data. Yet, LLMs aren't producing word-class stories as yet. However, there is certainly explicit knowledge about the craft of writing stories that has been shared by literary scholars as well as writers, especially in teaching environments. There are very many prescriptions for best practices of how to write well, many of them varying with the genre and type of writing task, and some clearly inconsistent with each other. For example, here is a handful of popular

6.4 Contemporary Narrative Generation Trends

writing maxims.[5] Bear in mind that many of these are to be taken with a pinch of salt, and that some excellent writers thrive on violating them:

- "Show, Don't Tell". This maxim is immensely popular, but ignores the importance of diegesis, as I pointed out earlier in Chap. 3, Sect. 3.2.2. It is typically attributed to Chekhov's brilliant advice: "Don't tell me the moon is shining; show me the glint of light on broken glass".
- "Write What You Know" (Mark Twain).
- "Avoid Adverbs" (Stephen King).
- "Start with a Compelling Hook" (unknown).
- "Omit Needless Words" (from Strunk and White's *The Elements of Style*).
- "Use Active Voice" (Strunk and White).
- "Don't Start with the Weather" (Elmore Leonard).
- "Make Your Characters Want Something" (Kurt Vonnegut).
- "Never Use a Verb Other Than 'Said' to Carry Dialogue" (Elmore Leonard).
- "Write naked. That means to write what you would never say" (Denis Johnson).
- "Read over your compositions, and wherever you meet with a passage which you think is particularly fine, strike it out." This maxim, from Dr. Samuel Johnson, is one of my favorites.

If this sort of craft knowledge was collected together in textual form, that might be sufficient for a machine-learning algorithm to test against the vast corpora of stories that are out there. The AI system could then arrive at its own weights for particular best practices in different contexts.

One way to carry this out would be to turn a rule into a prompt (e.g., "Does the opening sentence start with a compelling hook?") and then use an LLM to apply that prompt to example texts and answer 'Yes' or 'No' or provide a score in each case. If humans had separately labeled the stories (say in existing online reviews) as good or bad, one could then train a machine-learning algorithm to predict which combinations of rules result in better stories. A paper by Smith et al. (2024) takes a similar approach for the general problem of using model prompts to label examples.

Such a trained 'Story Craft' system could be useful in co-authoring stories with machines, and seems worthy of research exploration.

---

[5] I will avoid formal citations for these maxims as their true origins are often disputable.

6.4.2 Evaluation Challenges

6.4.2.1 Introduction

Systems that generate NL output are often evaluated intrinsically in terms of quality and informativeness.[6] Quality usually involves human-assessed measures of fluency, coherence, etc., though these can be approximated by automatic methods. Coherence is actually a many-faceted concept that has not been fleshed out adequately for narrative, as we shall see.

Informativeness has to do with how well the output preserves information content, which is a well-defined notion in the case of machine translation (where it is sometimes called adequacy) and in summarization, where source texts are being mapped to target ones. It is more easily computed against human-created reference texts (translations and summaries), leading to easily computable text similarity metrics, especially n-gram level comparison metrics. The fact that multiple informative and fluent summaries are possible for a given source, in other words, the absence of a unique correct answer, is partially addressed by using multiple reference translations or summaries. Automatic comparisons of system output against reference data can also involve syntactic, semantic, or discourse-level representations, when these can be computed accurately, as well as word embeddings and output activations.

However, such comparisons penalize the diversity of machine-generated output. Using neural approaches, word embeddings as well as output activations have been compared. While these automatic metrics often correlate with human judgments, they are not theoretically well-motivated. Nevertheless, they are of considerable value in large-scale evaluations, and some of these metrics can be used during training as well.

Many of these and similar automatic pairwise comparison metrics have been applied to NL generation. Here we want to reward diversity and creativity in a generated text, which might not overlap with any reference summary. Further, if the reference text itself lacks creativity, a metric that assigns a high score to the machine text compared with it may not be meaningful. Nevertheless, practical considerations have spawned a lot of interest in such automatic generation metrics. Surveying them here is outside the scope of the book. However, I would like to single out one metric for comparing texts: MAUVE, which examines overall patterns in word usage rather than focusing on each individual word. It can signal when a text is too repetitive or when it becomes overly random and loses coherence. The interested reader who wants to delve into the mathematical details can find them in the next section; others may prefer skipping it.

---

[6] These sorts of evaluations are different from extrinsic evaluations that measure performance in some task. While the latter is very useful for practical purposes, such evaluations are more complex to design, and don't lead to easily controllable experiments that yield scientific insights.

6.4.2.2 The MAUVE Metric: Details*

One relatively well-motivated family of metrics is to compare texts at the level of probability distributions. We already saw, in Chap. 1, metrics such as perplexity and KL-divergence could be used to compare a machine-learned model distribution Q to a true distribution P. In Chap. 2, we also discussed the work of Basu et al. (2021). They found that the hyper-parameter k, which is used by LLMs to select the top k most probable next words (or, in the case of nucleus sampling, the top $k\%$ of the probability mass) can cause *repetition* if k is too low, and *incoherence*—in a broad sense—if it's too high. Therefore, when comparing Q with P, it makes sense to try and balance between these two particular types of errors, repetition and incoherence, respectively. Using a sliding parameter λ to do just that, MAUVE (Pillutla et al. 2021) compares the machine-generated text distribution Q to a human-generated text distribution P using Eq. 6.1:

$$C(P, Q) = \{ (e^{-c\,KL(Q\|R_\lambda)}, e^{-c\,KL(P\|R_\lambda)}) : \\ R_\lambda = \lambda P + (1 - \lambda)Q, \text{ for } \lambda \in \{0, 1\} \} \quad (6.1)$$

> **Explanation of Eq. 6.1**
> C here represents a divergence curve. It considers the set of all pairs consisting of the negative exponent of the KL-divergence of $Q \parallel R_\lambda$ and the negative exponent of the KL-divergence of $P \parallel R_\lambda$, where R_λ, which is set by trial and error between zero and 1, is the slider mentioned earlier, and where little c is another hyper-parameter, greater than zero.

The MAUVE metric computes the area under the divergence curve C for pairs of texts submitted as input. Computed MAUVE scores for comparisons (of different machine text continuations given a prompt) show high correlations with human judgments of how human-like, interesting, and sensible the machine text is. Unfortunately, no comparisons against human-generated reference texts were carried out. It's also unclear that the metric rewards originality and creativity.

6.4.2.3 Repetition as a Goal

It's worth pointing out here that contrary to the dull repetition of a 'stuck' LLM, certain forms of repetition are rather desirable as a literary device. In traditional rhetoric (Quinn 1982), these include the ones in Example 6.13. Why do I list these? Because if Generative AI is ever to impress writers and literary scholars, story generators need to be able to invoke such devices at will to produce creative turns of phrase. Though many of the examples are from the literature of earlier centuries, modern equivalents are easily constructed.

Example 6.13

- **accumulatio**: (Repetition) "I will not excuse you you shall not be excus'd; excuses shall not be admitted".
- **anadiplosis**: (Repetition of end at next beginning) "When I give, I give myself."
- **antanaclasis**: (Repetition with different senses) "Let the dead bury their dead."
- **diacope**: (Repetition with only 1-2 words in between) "Villain, damned smiling villain."
- **epanados**: (Repetition in opposite order) "Fair is foul and foul is fair."
- **epanalepsis**: (Repetition of beginning at end) "Common sense is not so common."
- **epistrophe**: (Repetition of ends) "When I was a child, I spake as a child, I understood as a child, I thought as a child."
- **epizeuxis**: (Immediate repetition) "Comfort ye, comfort ye, my people."
- **gradatio**: (Repeated anadiplosis) "Tribulation worketh patience; and patience, experience; and experience, hope; and hope maketh not ashamed."
- **inclusio**: (Repetition of beginning of long passage at end) "What doth it profit, my brethren, though a man say he had faith . . . what doth it profit?"
- **isocolon**: (Repetition of grammatical forms) "The bigger they are, the harder they fall."
- **ploce**: (Repetition of word in general and then restricted sense) "more Irish than the Irish."
- **polyptoton**: (Repetition of same word in different grammatical forms) "Few men speak humbly of humility."
- **repetitio**: (Irregular repetition of word or phrase) "Light seeking light does light of light beguile."
- **symploce**: (Repetition of both beginnings and endings) "Are they Hebrews? So am I. Are they Israelites? So am I."

Can today's narrative generation tools actually produce these sorts of creative repetitions? Given Example 6.13 and our ROC spaghetti story, Example 6.14 is what GPT-4o produces:

Example 6.14

> **User Prompt**: Here are some forms of literary repetition . . .
> Please rewrite the following text to use all these forms of repetition as possible, keeping the same length. Indicate in parentheses after each sentence the name of these devices used. . . .
> **Machine**: Tina made spaghetti for her boyfriend. Spaghetti that took work, hard work—work that required time, patience, and effort. (epizeuxis, isocolon)
> She was proud, proud beyond measure, proud of her cooking, proud of herself. (epizeuxis, repetitio)
> Her boyfriend ate the whole plate, yes, he ate and ate, clearing the plate and saying it was good. (diacope, repetitio)

> When she tasted it herself, she tasted bitterness, a bitterness she didn't expect. (anadiplosis, polyptoton)
> The flavor was foul—foul in every way. (epanados)
> Yet, he smiled as if it were good, good for her, if not good for him. (antanaclasis)
> And in that moment, she was touched. Touched by his kindness, touched by his care. (epizeuxis, diacope)
> Her heart, her caring heart, warmed, realizing that the spaghetti wasn't good, but he made it good. (diacope, ploce)
> What does it mean, she thought, to love like this—what does it mean? (inclusio)
> It was both a good meal and a bad one. Good for her, not for him. (antanaclasis, epanalepsis)
> But he smiled, and smiled, as if it were good. (epistrophe)
> The more she thought, the more she felt gratitude; and gratitude brought her joy, and joy brought her peace. (gradatio)
> After all, if he could love her food, maybe he could love her, too. Love her as she loved him. (symploce)

The reader can verify, as the machine later does, how each of these literary devices is used. It can be seen that such creativity is not out of reach of the LLMs with their default settings for inference such as top-k, temperature, etc. Avoiding dull repetition but encouraging creative uses of it seems to depend on the quality of the prompting.

6.4.2.4 Desiderata for Generated Story Quality

The types of coherence analyses used for assessing the quality of NL generation, whether automatically or by human judges, can definitely be expanded based on the narratological insights provided in this book. Here are some of the new features that are relevant for generating the plot as well as the final text, that are also significant for evaluation:

- **Focalization**: We saw that different kinds of focalization and their mixed forms all contribute highly to the interpretation and aesthetics of narrative. Tracking focalization in terms of weight vectors should be possible.
- **Characterization**: Are the characters interesting, and believable? We saw in Chap. 3 statistical techniques for detecting characters and their attributes (defined in different ways) and inter-relationships. Leveraging this sort of information could provide more fine-grained and enhanced control over narration.
- **Tempo**: Does the narrative move along quickly, or does it dawdle? These perceptions of course depend on the stage of the narrative, and commonsense estimates of elapsed time in the story. As discussed in Chap. 5, tempo is measurable. Learning when to speed up or slow down would be extremely useful.

- **Narrative Distance**: This feature tracks shifts in narrative distance, e.g., changing from first to third, or from narrated speech to transposed speech to quotation. If the training data is not dominated by poorly constructed fictions, it's possible that abrupt shifts will be avoided at story generation time.
- **Narrative Order:** We have seen in considerable detail in Chap. 3 the various ways in which we can order narratives, and have found, in Sect. 6.3, that LLMs are quite proficient at that. Nevertheless, knowing when to deviate from chronological order is something that systems have not so far gotten a handle on. Skilled writers can often sense how much backstory is tolerable without distracting from the onward progression of the narrative. Changes in narrative order, which as we have seen in Chap. 5, are easily detectable by machines, can be useful to track.
- **Closure**: Is there a worthwhile ending? Endings are a distinguished element of any story, and a crucial determinant of the overall audience response. We are all familiar with cliffhanger, happy, sad, inconclusive, or open-ended endings, and even accomplished writers struggle with how to end a story (killing off the principal characters or antagonists is in general rather unsatisfying, though *Hamlet* is a notable exception). Interactive fiction provides a more constrained set of possibilities, especially those derived from games. In future, we need to examine character outcomes at the ends of narratives and develop methods to weight them.
- **Suspense**: Are expectations set up which engage the reader? Is foreshadowing and dramatic irony used effectively? As seen in Sect. 6.2.4.2, there has been interesting research on suspense for interactive narrative. In recent work (Xie et al. 2024), iterative zero-shot LLM prompting is used to generate suspenseful story plans, with human evaluations validating its effectiveness. The process involves outlining the protagonist, their goal, and a dire consequence for failure, followed by generating actions to achieve the goal and blocking conditions that cause potential failure, with detailed elaboration of event sequences. Developing suspense measures could help guide the use of such extensive prompts.
- **Surprisal**: The hypothesis of Uniform Information Density (UID), discussed in Chap. 1 holds that speakers will try minimize large jumps in surprisal across an utterance. For narrative generation, however, the surprisal has to vary at crucial points in the narrative to create unexpected twists and development. We have already mentioned the relationship of surprisal to dull repetition and incoherence in LLMs. Therefore, computing the surprisal throughout and training based on those weights, can be useful.
- **Causality**: Are there preconditions being met for the next sentence, given the postconditions from the previous one? We saw how to compute these logical relationships within LLMs using commonsense knowledge graphs, and so learning to optimize sequences of these relations during training can be valuable for a more coherent plot and output story.

6.5 NarrativeML Redux

The one thing I have left out is logical consistency. While causal relations address an aspect of it, a richer integration of logic within neurosymbolic computing is required to fully address that.

These features may be implemented in various ways, including additional embeddings with intermediate-layer integration into transformer blocks (the latter was mentioned when we discussed RAG methods in Chap. 2, Sect. 2.7.2).

6.5 NarrativeML Redux

To address the specific needs of story generation that haven't been captured so far, we now add plot-related preconditions and postconditions to events. The exact logic within the conditions can be any formula (and is thus left open as arbitrary character data in XML).

Example 6.15

```
<?xml version="1.0" encoding="UTF-8"?>
<!DOCTYPE NarrativeML [
<!DOCTYPE NarrativeML [
...
<!ELEMENT NARRATIVE (#PCDATA | NARRATOR |
AUDIENCE | CHARACTER |
 EVENT | TIME | TLINK | SLINK |
 EVALUATION |
 GOAL | TEMPO | NEC | PLOT | CONDITION |
 SEGMENT)*>
 ...
<!ELEMENT PLOT EMPTY>
<!ATTLIST PLOT id ID #REQUIRED>
<!ATTLIST PLOT NECS IDREFS #IMPLIED>
<!ATTLIST PLOT GOALS IDREFS #IMPLIED>
...
<!ATTLIST EVENT CONDITIONS IDREFS #IMPLIED>
...
<!ELEMENT CONDITION EMPTY>
<!ATTLIST CONDITION id ID #REQUIRED>
<!ATTLIST CONDITION event IDREF #REQUIRED>
<!ATTLIST CONDITION logic CDATA #REQUIRED>
<!ATTLIST CONDITION type (pre | post) #REQUIRED>
...
]>
```

Figures 6.1 and 6.2 shows the human-verified NarrativeML for our spaghetti story. The NarrativeML is broken into two parts to make things clearer. The reader can track through the goal tree in Fig. 6.2 to see how the character goals are laid out.

```xml
<NarrativeML version="2.2">
<NARRATIVE id="i1" level="0">
<NARRATOR id="n1" name="Narrator" person="3sg" exists="true"
        distance="NARRATED" type="absent"
        order="CHRONICLE" timeRelation="SUBSEQUENT"
              perspective="INTERNALLY_FOCALIZED"
        speechTime="t1" />
<CHARACTER id="c1" name="Tina" type="animate" exists="true"
        accessible_to="c1 c3" />
<CHARACTER id="c2" name="spaghetti" type="inanimate" exists="true"
        accessible_to="c1 c3" />
<CHARACTER id="c3" name="boyfriend" type="animate" exists="true"
        accessible_to="c1 c3" />
<EVENT id="e1" participants="c1 c2">made</EVENT>
<EVENT id="e2" participants="c1">proud</EVENT>
<EVENT id="e3" participants="c3 c2">ate</EVENT>
<EVENT id="e4" participants="c3">said</EVENT>
<EVENT id="e5" participants="c1 c2">tried</EVENT>
<EVENT id="e6" participants="c1">realized</EVENT>
<EVENT id="e7" participants="c1">touched</EVENT>
<EVENT id="e8" participants="c3">pretended</EVENT>

<TLINK id="tr1" type="BEFORE" eventID="e1" relatedToEvent="e2" />
<TLINK id="tr2" type="BEFORE" eventID="e2" relatedToEvent="e3" />
<TLINK id="tr3" type="BEFORE" eventID="e3" relatedToEvent="e4" />
<TLINK id="tr4" type="BEFORE" eventID="e4" relatedToEvent="e5" />
<TLINK id="tr5" type="BEFORE" eventID="e5" relatedToEvent="e6" />
<TLINK id="tr6" type="SIMULTANEOUS" eventID="e6" relatedToEvent="e7" />
<TLINK id="tr7" type="SIMULTANEOUS" eventID="e8" relatedToEvent="e4" />

<NEC id="nec1" entity="c1" events="e1 e2 e5 e6 e7" />
<NEC id="nec3" entity="c2" events="e1 e3 e5" />
<NEC id="nec2" entity="c3" events="e2 e3" />

<TEMPO storyTime="PT2H30M" readingLength="PT45S" />
<EVALUATION id="ev1" eventID="e4" characterID="c3" audienceID="reader1"
        value="1" polarity="positive" />
<EVALUATION id="ev2" eventID="e6" characterID="c1" audienceID="reader1"
         value="1" polarity="negative" />
<EVALUATION id="ev3" eventID="e7" characterID="c1" audienceID="reader1"
        value="1" polarity="positive" />
</NARRATIVE>
</NarrativeML>
```

Fig. 6.1 Human-verified NarrativeML for spaghetti story (Part I)

6.5 NarrativeML Redux

```xml
<NarrativeML version="2.2">
  <NARRATIVE id="i1" level="0">
    ...

<EVENT id="e1" participants="c1 c2">made</EVENT>
<CONDITION id="cond1" event="e1" type="pre"
        logic="Exists(x) (Ingredient(x) AND Has(c1, x))" />
<CONDITION id="cond2" event="e1" type="post"
           logic="Cooked(c2)" />

<EVENT id="e2" participants="c1">proud</EVENT>
<CONDITION id="cond3" event="e2" type="pre"
           logic="Cooked(c2)" />
<CONDITION id="cond4" event="e2" type="post" logic="Proud(c1, e1)" />

<EVENT id="e3" participants="c3 c2">ate</EVENT>
<CONDITION id="cond5" event="e3" type="pre" logic="OnPlate(c2)" />
<CONDITION id="cond6" event="e3" type="post" logic="not(OnPlate(c2))" />

<EVENT id="e4" participants="c3">said</EVENT>
<CONDITION id="cond7" event="e4" type="pre" logic="Ate(c3, c2)" />
<CONDITION id="cond8" event="e4" type="post" logic="Believe(c3, Good(c2))" />

<EVENT id="e5" participants="c1 c2">tried</EVENT>
  <CONDITION id="cond9" event="e5" type="pre" logic="not(Empty(c2))" />
  <CONDITION id="cond10" event="e5" type="post"  logic="Tasted(c1, c2)" />

<EVENT id="e6" participants="c1">realized</EVENT>
<CONDITION id="cond11" event="e6" type="pre" logic="Tasted(c1, c2)" />
<CONDITION id="cond12" event="e6" type="post" logic="Aware(c1, not(Good(c2)))" />

<EVENT id="e7" participants="c1">touched</EVENT>
<CONDITION id="cond13" event="e7" type="pre" logic="Complimented(c3, c2)" />
<CONDITION id="cond14" event="e7" type="post"
           logic="Touched(c1, Complimented(c3, c2))" />

<EVENT id="e8" participants="c3">pretended</EVENT>
<CONDITION id="cond15" event="e8" type="pre" logic="Aware(c3, not(Good(c2)))" />
<CONDITION id="cond16" event="e8" type="post" logic="Pretended(c3, Good(c2)) />

<GOAL id="g1" parent="g2" character="c1" leaf="true" events="e1">cook</GOAL>

<GOAL id="g2" parent="" character="c1" leaf="false">show love</GOAL>

<GOAL id="g3" parent="g2" character="c1" leaf="true" events="e2">feel proud</GOAL>

<GOAL id="g4" parent="" character="c2" leaf="false">show appreciation</GOAL>

<GOAL id="g5" parent="g4" character="c2" leaf="true" events="e4">say</GOAL>
</NARRATIVE>
</NarrativeML>
```

Fig. 6.2 Human-verified NarrativeML for spaghetti story (Part II)

6.6 Digression: NarrativeML Plot-level Analysis

Now that we have the plot fleshed out in NarrativeML, let us make sure LLMs can analyze stories in terms of Plot.

Let's give the LLM the above NarrativeML DTD and an example of the San Cristobal de las Casas biking opening sentence from Chap. 5 (Example 5.33), along with its machine-generated NarrativeML output shown there (which has no plot information). Then, in this one-shot exercise, let's feed the LLM our ROC spaghetti story (Mostafazadeh et al. 2016) as input, asking it to fill out the plot:

Example 6.16

> **User Prompt**: Here is the NarrativeML DTD, and a sample story and its annotation ... Please generate the NarrativeML for this new story ... Also fill out the plot, with the character goals and logic for the preconditions and postconditions for each event specified in predicate calculus:
>
> Tina made spaghetti for her boyfriend. It took a lot of work, but she was very proud. Her boyfriend ate the whole plate and said it was good. Tina tried it herself, and realized it was disgusting. She was touched that he pretended it was good to spare her feelings.

Figure 6.3 shows the output from GPT-4o (re-formatted slightly). The event extents continue to be too extensive, but the event types are correct. Regarding the temporal structure, the TLINKS are all correct, and the time duration estimates, which we discovered that humans find difficult, are plausible. The character attributes aren't filled out. The character evaluations are incorrect in that they are by characters, rather than by the audience, but the polarity is right. Regarding the plot, the character goals lack the proper hierarchy and the events are linked to them, but the preconditions and postconditions for the plot are filled out automatically, and correctly in predicate calculus except for 'not(Exists(c3, Plate))', which should be 'not(OnPlate(c3))'. The extraction of logic from a story, without any prompting examples, is truly a demonstration of emergent commonsense knowledge based on the rich contextual approach to meaning we discussed in Chap. 2—rather than any sort of stochastic parroting or mimicry.

6.7 NarrativeML Story Generation

6.7.1 Using Noisy NarrativeML as Fabula

Now, let's use another LLM to generate a story based on machine-generated NarrativeML. Example 6.17 shows the (zero-shot) output from Claude 3.5 Sonnet, after feeding it the NarrativeML DTD and, as input, the unedited machine NarrativeML output for our spaghetti story, which was shown above in Fig. 6.3:

6.7 NarrativeML Story Generation

```
<NarrativeML version="2.0">
<NARRATIVE id="n1" title="Tina's Spaghetti" medium="descriptive_text">
<NARRATOR id="n2" type="present" coref="c1" exists="true"
          order="CHRONICLE" distance="DIRECT"
          perspective="INTERNALLY_FOCALIZED" timeRelation="SUBSEQUENT" />
<CHARACTER id="c1" name="Tina" type="animate" exists="true" accessibleTo="c1" />
<CHARACTER id="c2" name="Boyfriend" type="animate" exists="true" accessibleTo="c1" />
<CHARACTER id="c3" name="Spaghetti" type="inanimate" exists="true" accessibleTo="c1" />

<EVENT id="e1" type="ACTION" participants="c1 c3" duration="PT2H">make spaghetti</EVENT>
<CONDITION id="cond1" event="e1" type="pre" logic="Exists(x) (Ingredient(x) AND Has(c1, x))" />
<CONDITION id="cond2" event="e1" type="post" logic="Cooked(c3)" />

<EVENT id="e2" type="MENTAL" participants="c1" duration="PT1M">feel proud</EVENT>
<CONDITION id="cond3" event="e2" type="pre" logic="Cooked(c3)" />
<CONDITION id="cond4" event="e2" type="post" logic="Proud(c1, e1)" />

<EVENT id="e3" type="ACTION" participants="c2 c3" duration="PT15M">eat spaghetti</EVENT>
<CONDITION id="cond5" event="e3" type="pre" logic="OnPlate(c3)" />
<CONDITION id="cond6" event="e3" type="post" logic="not(Exists(c3, Plate))" />

<EVENT id="e4" type="SPEECHACT" participants="c2" duration="PT1M">say it was good</EVENT>
<CONDITION id="cond7" event="e4" type="pre" logic="Ate(c2, c3)" />
<CONDITION id="cond8" event="e4" type="post" logic="Believe(c1, Good(c3))" />

<EVENT id="e5" type="PERCEPTION" participants="c1" duration="PT2M">try spaghetti</EVENT>
<CONDITION id="cond9" event="e5" type="pre" logic="not(Empty(c3))" />
<CONDITION id="cond10" event="e5" type="post" logic="Tasted(c1, c3)" />

<EVENT id="e6" type="MENTAL" participants="c1" duration="PT1M">realize it was disgusting</EVENT>
<CONDITION id="cond11" event="e6" type="pre" logic="Tasted(c1, c3)" />
<CONDITION id="cond12" event="e6" type="post" logic="Aware(c1, not(Good(c3)))" />

<EVENT id="e7" type="MENTAL" participants="c1" duration="PT1M">feel touched by boyfriend's kindness</EVENT>
<CONDITION id="cond13" event="e7" type="pre" logic="Complimented(c2, c3)" />
<CONDITION id="cond14" event="e7" type="post" logic="Touched(c1, Compliment(c2, c3))" />

<TLINK id="tr1" type="BEFORE" eventID="e2" relatedToEvent="e3" />
<TLINK id="tr2" type="BEFORE" eventID="e3" relatedToEvent="e5" />
<TLINK id="tr3" type="SIMULTANEOUS" eventID="e6" relatedToEvent="e7" />

<NEC id="nec1" entity="c1" events="e1 e2 e5 e6 e7" />
<NEC id="nec2" entity="c2" events="e3 e4" />
<NEC id="nec3" entity="c3" events="e1 e3 e5" />

<EVALUATION id="ev1" eventID="e4" characterID="c2" audienceID="c1" value="1" polarity="positive" />
<EVALUATION id="ev2" eventID="e6" characterID="c1" audienceID="c1" value="1" polarity="negative" />

<GOAL id="g1" parent="p1" character="c1" leaf="true">make a meal for her boyfriend</GOAL>
<GOAL id="g2" parent="g1" character="c1" leaf="true">impress her boyfriend</GOAL>
<GOAL id="g3" parent="g1" character="c1" leaf="true">feel proud of her cooking</GOAL>
<GOAL id="g4" parent="g2" character="c2" leaf="true">show appreciation for Tina?s effort</GOAL>

<TEMPO storyTime="PT2H30M" readingLength="PT45S" />
</NARRATIVE>
</NarrativeML>
```

Fig. 6.3 Machine-generated NarrativeML for prompt in Example 6.16

Example 6.17

> **User Prompt**: Here is the NarrativeML DTD ... and a markup for a potential story in NarrativeML ... Please generate a five-sentence story based on it. ...
> **Machine**: *Tina spent two hours carefully preparing spaghetti for her boyfriend, feeling proud of her culinary creation as she set it on the table. Her boyfriend ate the meal enthusiastically for fifteen minutes, offering genuine praise and telling her how good it was. After he finished, Tina finally had a chance to try some of the remaining spaghetti herself, excited to taste her successful dish. To her horror, she realized the spaghetti was absolutely disgusting—she had somehow confused salt and sugar while cooking.*

> Yet in that same moment, as she processed both the awful taste and her boyfriend's earlier compliments, she felt deeply touched by his kindness in eating the entire portion without a single complaint.

We can see from this that even noisy NarrativeML is expressive enough for story generation. How about trying an entirely different LLM? Here is the (zero-shot) output from Hugging Face's interface to the Meta/Facebook model Meta-Llama-3.1-70B-Instruct, after feeding it the NarrativeML DTD and, as input, the machine output in Fig. 6.3 (in the latter, I changed Tina to Mary and spaghetti to chicken curry, just in case the LLM was despite clearing memory already familiar with the story):

Example 6.18

> **User Prompt**: Here is a markup for a potential story in NarrativeML. ... Please generate a five-sentence story based on it. ...
> **Machine**: *Mary spent two hours cooking a delicious chicken curry for her boyfriend. When it was finally ready, she felt a surge of pride and couldn't wait for him to try it. Her boyfriend took a bite and exclaimed that it was good, which made Mary feel happy and touched by his kindness. However, when Mary tried the curry herself, she was shocked to realize that it was actually disgusting. Despite the disappointment, Mary felt grateful for her boyfriend's thoughtful gesture and appreciated the effort he put into making her feel special.*

From these examples, it is clear that NarrativeML is sufficiently detailed for generating a story.

6.7.2 Leaving Plot Out of Fabula

It's possible to leave out the plot, corresponding to surface generation. Let's feed an LLM the NarrativeML DTD and, as input, the machine output in Example 6.16 with the plot and conditions left out (here I changed Tina to Parvati and spaghetti to molaghuttal, which is a somewhat obscure South Indian dish). Here is the (zero-shot) output from Hugging Face's interface to the Cohere model c4ai-command-r-plus-08-2024:

6.7 NarrativeML Story Generation

Example 6.19

> **User Prompt**: Here is a markup for a potential story in NarrativeML. . . . Please generate a five-sentence story based on it. . . .
> **Machine**: *Parvati, a skilled cook, spent two hours in the kitchen preparing her signature dish, molaghuttal, a traditional Indian stew. She took great pride in her culinary creation, feeling a sense of accomplishment as she plated the vibrant dish. Her boyfriend, eager to taste her cooking, sat down and devoured the molaghuttal within fifteen minutes, a testament to his appreciation. After he finished, he praised her cooking, saying it was delicious. However, Parvati, being a perfectionist, decided to try the dish herself and realized, to her surprise, that it was not as tasty as she had expected.*

This model is less sophisticated, in that it doesn't manage to get to Parvati's appreciation for her boyfriend's gesture, undoubtedly because it ran out of space, as I verified by increasing the length to six sentences—by which time it does get to it. Once again, we see that the lack of a knapsack approach to generation results in an inability to adhere to length constraints.

6.7.3 Providing Partial Plot in Fabula

It is far more interesting to give the machine a partial plot, leaving out the details of the events, and have it complete the story. Let's pass in as the plot just the first three events with their conditions in predicate logic from the machine output in Fig. 6.3, but without the event tag details. Here I deliberately edited out the character goals, but gave the principal character some attributes. For fun, I also made the dish really exotic, and changed the proper name of the character:

Here is the (zero-shot) output by GPT-4o after giving it the NarrativeML DTD, and feeding it the initial NarrativeML shown in Fig. 6.4.

Example 6.20

> **User Prompt**: Here is an input NarrativeML that specifies the initial part of the story. . . . Generate a five-sentence completed story for it.
> **Machine**: *Numm carefully prepared the ingredients for the balut, drawing on her cooking skills to create something special. Once the balut was cooked to perfection, she felt a deep sense of pride in her work, eager to share it. She placed the balut on a plate and offered it to her boyfriend, who accepted it with a smile. After watching him enjoy the meal, Numm decided to try it herself, only to realize that it tasted unusual, not quite what she expected. Despite her disappointment, she was touched to see that her*

> *boyfriend had finished the meal without a word of complaint, showing his appreciation for her effort and kindness.*

The style is over-the-top, which of course can be controlled somewhat by enhancing the prompt. The main takeaway from this example is to illustrate that it's viable to use NarrativeML plots at the level of predicate logic to generate stories using LLMs.

All in all, we can conclude that NarrativeML is fine-grained and expressive enough for story generation.

6.8 Status of Story Generation Systems

Looking over story generation systems as a whole, there are several observations to be made:

- *Narrative Structure*: Generators don't do well on structuring anything other than short spans of text, even when they avail of high-level character and narrative goals. Here they stand to benefit from the narratological distinctions described in Chap. 3, which can of

```xml
<?xml version="1.0" encoding="UTF-8"?>
<!DOCTYPE NarrativeML SYSTEM "NarrativeML.dtd">
<NarrativeML version="2.0">
  <NARRATIVE id="n1" title="Numm's balut" medium="descriptive_text">
    <NARRATOR id="n2" type="present" coref="c1" exists="true"
      order="CHRONICLE" distance="DIRECT" perspective="INTERNALLY_FOCALIZED"
      timeRelation="SUBSEQUENT"/>
    <CHARACTER id="c1" name="Numm" type="animate" exists="true"
      accessibleTo="c1" attributes="kind, skilled cook"/>
    <CHARACTER id="c2" name="Boyfriend" type="animate" exists="true"
      accessibleTo="c1"/>
    <CHARACTER id="c3" name="balut" type="inanimate" exists="true"
      accessibleTo="c1"/>
      <SEGMENT id="s1" title="">
        <CONDITION id="cond1" event="e1" type="pre"
          logic="Exists(x)(Ingredient(x) &amp; Has(c1, x))"/>
        <CONDITION id="cond2" event="e1" type="post" logic="Cooked(c3)"/>
        <CONDITION id="cond3" event="e2" type="pre" logic="Cooked(c3)"/>
        <CONDITION id="cond4" event="e2" type="post" logic="Proud(c1, e1)"/>
        <CONDITION id="cond5" event="e3" type="pre" logic="OnPlate(c3)"/>
        <TLINK id="tr1" type="BEFORE" eventID="e2" relatedToEvent="e3"/>
      </SEGMENT>
      <PLOT id="p1" NECS="nec1 nec2 nec3"/>
      <NEC id="nec1" entity="c1" events="e1 e2"/>
      <NEC id="nec2" entity="c2" events="e3"/>
      <NEC id="nec3" entity="c3" events="e1 e3"/>
      <TEMPO storyTime="PT2H30M" readingLength="PT45S"/>
  </NARRATIVE>
</NarrativeML>
```

Fig. 6.4 Initial GPT-4o + lightly human-edited NarrativeML for Example 6.16.

course be specified in NarrativeML. However, as we have seen, current LLM technology is not that useful for story generation as incoherence arises quickly, and long-distance narrative dependencies aren't addressed. Accordingly, bespoke implementations may work better.

- *Tradition*: Even the most readable of generated stories is far different from what readers are used to. Readers are used to placing human-written stories in a tradition or context that comes with cultural knowledge, including prior reading. Often, upon reading the opening sentences of such a story, the reader may know what sort of story they are reading. A reader today might detect influences from magical realism, minimalism, hardboiled detective story, Lovecraftian horror, Gothic fiction, sword-and-sorcery, cyberpunk, hard science-fiction, etc. (These are not usually hard-and-fast categories into which fiction should be classified, but softer ones that are perhaps better modeled through weights and clusters of similar stories, and perhaps tie in with the Bayesian model of reading discussed in Chap. 4). Readers are also likely to have different prior expectations if they know a story is machine-generated. Rather than trying to hide the fact, machine-generated stories are best analyzed in terms of their history, as revealed in the compendium by Bertram and Montfort (2024), which includes research systems as well as the work of independent story programmers.

- *Piggybacking*: Some of that history include stories that are variations of human-generated ones, like the 2014 story by David Stark called *Moebius Tentacle; or the Space-Octopus* (see Bertram and Montfort 2024, p. 74), where words in *Moby Dick* are substituted with an entertaining result. These tend to work better than others because they rely on the parent from which it is grafted for structure. To go beyond mimicry, however, the variations must be diverse and random enough to convey spontaneity and effect surprise.

- *Text for Gaming*: Computer game fictions are generally more successful as they are a well-established contemporary idiom, as in the texts in the well-known Tarn and Zach Adams game of Dwarf Fortress (see Bertram and Montfort 2024, p. 384).

- *Constrained Writing*: Programs are a natural way of implementing highly constrained writing, much of it stemming from the mathematical and literary approach of the French Oulipo group,[7] as in the palindromic story *2002* (itself a palindromic calendar year) programmed by Montfort and Gillespie.[8] For that reason, computer-generated poetic forms that use well-established forms such as the sonnet, haiku, etc. seem more successful generally than prose.

---

[7] https://en.wikipedia.org/wiki/Oulipo.
[8] https://spinelessbooks.com/2002/.

6.9 The Hand of the Creator

Scientific publications these days often require authors to provide any code used in experiments on GitHub or elsewhere, so that results of the paper can be replicated as part of peer review. While this code is sometimes neglected by reviewers, in the case of machine-generated stories, it makes sense to provide the source code as it can be regarded as part of the creative artifact that includes the output.

The field of Critical Code studies (Marino 2020) treats computer code as a literary artifact like any other, carrying out a close reading of it. The goal is not to arrive at authorial intent or the meaning of the text, but an interpretation that focuses on the artifact and its significance within the relevant cultural milieu (or milieus). For example, *Taroko Gorge* is a poetry generator developed by Nick Montfort in 2009 and subsequently extended by others. The code (admirably—and perhaps even provocatively—concise) has been analyzed in Critical Code fashion by Marino (2020).[9] Marino notes that Montfort includes as a comment the location of where he wrote the code, meaning that he wanted readers to consider the composition location. It could, in my opinion, also be a deliberate borrowing from literary prefaces. The fact that Montfort uses many one-letter variable and function names suggests, according to Marino (2020), Montfort's wanting to give the code a minimalist feel. The code also uses only two tiny wordlists, of 8 words each, to create unbounded outputs.

If the code that produced the story output is treated as a first-class object of literary analysis, the hand of the creator need not be hidden but can be part and parcel of the creation itself. This implies, in turn, a view of computer-generated fiction as 'post-modernist', as we mentioned in passing in Chap. 3, in the sense of laying bare the scaffolding and processes underlying the creation of the literary artifact.

Of course, with LLM-based story generation, we are usually familiar with the architecture, and expect examples of prompts and API interactions to be provided. However, the LLM code code is hardly that revealing, compared to bespoke code that is custom-built for particular story generation tasks. When such code is generic and domain-independent, it is especially welcome, as in the case of *Taroko Gorge* above.

6.10 Conclusion

Initially, story generation went beyond scripts to include plans and case-based reasoning to improve structure and character representations. Early models focused on character goals and plans, evolving to also use narrator goals. Coherence was a clear challenge throughout the history of story generation. Interactive narrative planning can allow for immersive expe-

---

[9] For the *Taroko Gorge* demo and javascript source, point your web browser to https://collection.eliterature.org/3/files/taroko-gorge/taroko-gorge.html. For the (rather concise) Python code, visit https://nickm.com/code/taroko_gorge.py. For the poem's code critique by Marino (2020), see https://criticalcodestudies.com/tarokogorge.html.

riences, while allowing audiences limited control over outcomes. However, balancing authorial and audience control is non-trivial. Temporal generation systems can handle the seven types of narrative ordering, though knowing when to select a non-chronological ordering is still difficult. Current approaches to story generation primarily use neural models, where LLMs can produce short micro-fictions but struggle with coherence and length constraints, and especially so in longer-narratives. System-generated prompts offer enhanced control, as shown in studies that auto-generate prompts based on causal dependencies or learn plots from story datasets. Commonsense knowledge integration through neurosymbolic methods is also advancing.

For evaluation, intrinsic metrics focus on informativeness and quality, though rewarding diversity (including originality and creativity) remains challenging, especially when reference texts lack creativity. Metrics comparing distributional properties of generated and reference stories are among the more robust automatic evaluation methods. I proposed a set of key evaluation features for narrative, that can be incorporated into LLMs (using, say, intermediate-layer interleaving): focalization, characterization, tempo, narrative distance, order, closure, suspense, surprise, and causality.

Finally, plot-related preconditions and postconditions were added to NarrativeML. Analyzing micro-fictions using NarrativeML proves to be well within the capabilities of LLMs, including the ability to extract plot-related information with event preconditions and postconditions in first-order logic. I provided several convincing examples that suggest that NarrativeML is expressive enough to use as a basis for story generation with LLMs, whether at the level of specifying plot or specifying the actual events. I then offered an overall set of observations on the status of story generation systems today. Finally, I discussed some creative story generation efforts along with the relevance of Critical Code studies of the code used to generate the output. Further extensions, along with future prospects for story generation, are discussed in Chap. 7.

References

Aarseth, Espen J. (1997). *Cybertext: Perspectives on Ergodic Literature*. Baltimore: Johns Hopkins University Press.
Basu, S., G. S. Ramachandran, N. S. Keskar, and L. R. Varshney (2021). "Mirostat: A neural text decoding algorithm that directly controls perplexity". In: *International Conference on Learning Representations*. https://openreview.net/forum?id=W1G1JZEIy5%5C_.
Bertram, Lillian-Yvonne and Nick Montfort, eds. (2024). *Output: An Anthology of Computer-Generated Text, 1953-2023*. Cambridge, MA: MIT Press.
Borgeaud, Sebastian et al. (2022). "Improving Language Models by Retrieving from Trillions of Tokens". In: *Proceedings of the 39th International Conference on Machine Learning,* Baltimore, Maryland. https://proceedings.mlr.press/v162/borgeaud22a/borgeaud22a.pdf.

Bosselut, Antoine et al. (2019). "COMET: Commonsense Transformers for Automatic Knowledge Graph Construction". In: *Proceedings of the 57th Annual Meeting of the Association for Computational Linguistics (ACL'2019)*, pp. 4762–4779. https://aclanthology.org/P19-1470/.

Callaway, Charles (2000). "Narrative Prose Generation". PhD thesis. Department of Computer Science, North Carolina State University, Raleigh.

Cheong, Yun-Gyung (2007). "A Computational Model of Narrative Generation for Suspense". PhD thesis. Department of Computer Science, North Carolina State University.

Cheong, Yun-Gyung and R. Michael Young (2006). "A computational model of narrative generation for suspense." In: *Proceedings of the Association for Advancement of Artificial Intelligence (AAAI-2006) Computational Aesthetics Workshop*. Washington, DC: AAAI Press. https://cdn.aaai.org/AAAI/2006/AAAI06-331.pdf.

Fan, Angela, Mike Lewis, and Yann Dauphin (2019). "Strategies for structuring story generation". In: *Proceedings of the 57th Annual Meeting of the Association for Computational Linguistics (ACL'2019)*, pp. 2650–2660. https://aclanthology.org/P19-1254/.

Finnegan, Ruth (1977). *Oral Poetry: Its Nature, Significance and Social Context*. Bloomington, Indiana: Indiana University Press.

Genette, Gerard (1980). *Narrative Discourse*. Ithaca: Cornell University Press.

Gerrig, Richard and Dana Bernardo (1994). "Readers as problem-solvers in the experience of suspense". In: *Poetics* 22.6, pp. 459–472. http://dx.doi.org/10.1016/0304-422X(94)%2090021-3.

Gervas, Pablo (2014). "Story generator algorithms". In: *Handbook of Narratology*. Ed. by Peter Huhn, Jan Christoph Meister, John Pier, and Wolf Schmid. Berlin: De Gruyter, pp. 825–835.

Grasbon, Dieter and Norbert Braun (2001). "A morphological approach to interactive storytelling." In: *Proceedings of Artificial Intelligence and Interactive Entertainment (CAST'01), Living in Mixed Realities*. Sankt Augustin, Germany, pp. 337–340.

Hwang, Jena D. et al. (2021). "(COMET-)ATOMIC-2020: On Symbolic and Neural Commonsense Knowledge Graphs". In: *Proceedings of the AAAI Conference on Artificial Intelligence (AAAI-21)*, vol. 35. Washington, DC: AAAI Press, pp. 6384–6392. https://arxiv.org/abs/2010.05953.

Ippolito, Daphne, David Grangier, Chris Callison-Burch, and Douglas Eck (2019). "Unsupervised Hierarchical Story Infilling". In: *Proceedings of the First Workshop on Narrative Understanding*. https://aclanthology.org/W19-2405.

Kautz, Henry A. (1987). "A Formal Theory of Plan Recognition". PhD thesis. Dept. of Computer Science, University of Rochester.

Kautz, Henry A. and Bart Selman (1992). "Planning as satisfiability". In: *Proceedings of the 10th European Conference on Artificial Intelligence (ECAI'1992)*. Vienna, Austria, pp. 359–363. URL: https://www.cs.cornell.edu/selman/papers/pdf/92.ecai.satplan.pdf.

Kelly, Jack, Alex Calderwood, Noah Wardrip-Fruin, and Michael Mateas (2023). "There and Back Again: Extracting Formal Domains for Controllable Neurosymbolic Story Authoring". In: *Proceedings of the Nineteenth AAAI Conference on Artificial Intelligence and Interactive Digital Entertainment (AIIDE 2023)*. Washington, DC: AAAI Press. https://ojs.aaai.org/index.php/AIIDE/article/view/27502.

Kolodner, Janet (1993). *Extracting narrative timelines as temporal dependency structures*. San Mateo: Morgan Kaufmann.

Lebowitz, Michael (1985). "Story-telling as planning and learning". In: *Poetics* 14, pp. 483–502. URL: http://dx.doi.org/10.1016/0304-422X(85)%2090015-4.

Li, Boyang, Stephen Lee-Urban, George Johnston, and Mark O. Riedl (2013). "Story generation with crowdsourced plot graphs". In: *Proceedings of the Twenty-Seventh AAAI Conference on Artificial Intelligence (AAAI'13)*. Washington, DC: AAAI Press, pp. 598–604. https://ojs.aaai.org/index.php/AAAI/article/view/8649.

Lonneker, Birte (2005). "Narratological knowledge for natural language generation". In: *Proceedings of the 10th European Workshop on Natural Language Generation (ENLG'05)*. Aberdeen, Scotland, pp. 91–100. https://aclanthology.org/W05-1610/.

Lord, Albert (1960). *The Singer of Tales*. Cambridge, MA: Harvard University Press.

Loyall, A. Bryan and Joseph Bates (1991). *Hap: a reactive, adaptive architecture for agents*. Tech. rep. CMU-CS-91-147. Pittsburgh, PA: School of Computer Science, Carnegie Mellon University.

Marino, Mark C., ed. (2020). *Critical Code Studies*. Cambridge, MA: MIT Press.

Mateas, Michael and Andrew Stern (2002). "A behavior language for story-based believable agents." In: *Proceedings of the American Association for Artificial Intelligence (AAAI) Symposium on Artificial Intelligence and Interactive Entertainment*. Washington, DC: AAAI Press. URL: http://dx.doi.org/10.1109/MIS.2002.1024751.

Mateas, Michael and Andrew Stern (2005). "Structuring content in the Facade interactive drama architecture." In: *Proceedings of the First Conference on Artificial Intelligence and Interactive Digital Entertainment (AIIDE'2005)*. Washington, DC: AAAI Press, pp. 93–98. URL: https://ojs.aaai.org/index.php/AIIDE/article/view/18722.

McKee, Robert (1997). *Story: Substance, Structure, Style and the Principles of Screenwriting*. New York: Harper Collins.

Meehan, James R. (1976). "The Metanovel: Writing Stories by Computer". PhD thesis. Dept. of Computer Science, Yale University.

Montfort, Nick (2011). "Curveship's automatic narrative variation". In: *Proceedings of the 6th International Conference on the Foundations of Digital Games (FDG '11)*. Bordeaux, France, pp. 211–218.

Mostafazadeh, Nasrin, Nathanael Chambers, et al. (2016). "A corpus and evaluation framework for deeper understanding of commonsense stories". In: *Proceedings of the North American Chapter of the Association for Computational Linguistics: Human Language Technologies (NAACL-HLT'2016)*, pp. 839–849. https://aclanthology.org/N16-1098.pdf.

Murray, Janet H. (1997). *Hamlet on the Holodeck: The Future of Narrative in Cyberspace*. New York: The Free Press.

Peinado, Federico and Pablo Gervas (2006). "Evaluation of automatic generation of basic stories". In: *New Generation Computing* 24, pp. 289–302. URL: http://dx.doi.org/10.1007/BF03037336.

Peng, Nanyun, Marjan Ghazvininejad, Jonathan May, and Kevin Knight (2018). "Towards Controllable Story Generation". In: *Proceedings of the First Workshop on Storytelling, New Orleans*, pp. 43–49. https://aclanthology.org/W18-1505/.

Peng, Xiangyu, Siyan Li, Sarah Wiegreffe, and Mark Riedl (2022). "Inferring the Reader: Guiding Automated Story Generation with Commonsense Reasoning". In: *Findings of the Association for Computational Linguistics (EMNLP 2022)*, pp. 7008–7029. https://aclanthology.org/2022.findings-emnlp.520/.

Perez y Perez, Rafael and Mike Sharples (2004). "Three computer-based models of storytelling: BRUTUS, MINSTREL and MEXICA". In: *Knowledge-Based Systems* 17.1, pp. 15–29. URL: http://dx.doi.org/10.1016/S0950-7051(03)%2000048-0.

Pillutla, Krishna et al. (2021). "MAUVE: Measuring the Gap Between Neural Text and Human Text using Divergence Frontiers". In: *Proceedings of the 35th Conference on Neural Information Processing Systems (NeurIPS 2021)*. https://arxiv.org/abs/2102.01454.

Propp, Vladimir (1968). *Morphology of the Folktale*. Austin: University of Texas Press.U

Quinn, Arthur (1982). *Figures of Speech*. Layton, Utah: Gibbs Smith.

Riedl, Mark and Michael Young (2010). "Narrative planning: balancing plot and character." In: *Journal of Artificial Intelligence Research* 39, pp. 217–268. URL: http://dx.doi.org/10.1613/jair.2989.

Schank, Roger C. and Robert P. Abelson (1977). *Scripts, Plans, Goals and Understanding: An Inquiry into Human Knowledge Structures*. Hillsdale, NJ: Lawrence Erlbaum.

Scheidel, Antonia and Thierry Declerck (2010). "APftML – Augmented Proppian fairy tale Markup Language". In: *First International AMICUS Workshop on Automated Motif Discovery in Cultural Heritage and Scientific Communication Texts, Hungary*. Szeged University, Szeged. URL: https://www.dfki.de/fileadmin/user_upload/import/5050_AMICUS-APftML.pdf.

Smith, Ryan, Jason A. Fries, Braden Hancock, and Stephen H. Bach (2024). "Language Models in the Loop: Incorporating Prompting into Weak Supervision". In: 1.2. https://doi.org/10.1145/3617130.

Turner, Scott R. (1993). "Minstrel: A Computer Model of Creativity and Storytelling". PhD thesis. University of California at Los Angeles.

Wardrip-Fruin, Noah (2009). *Expressive Processing: Digital Fictions, Computer Games and Software Studies*. Cambridge, MA: MIT Press.

Xie, Kaige and Mark Riedl (2024). "Creating Suspenseful Stories: Iterative Planning with Large Language Models". In: *Proceedings of the 18th Conference of the European Chapter of the Association for Computational Linguistics (EACL'2024)*, pp. 2391–2407. https://aclanthology.org/2024.eacl-long.147/.

Yao, Lili et al. (2019). "Plan-And-Write: Towards Better Automatic Storytelling". In: *Proceedings of the AAAI Conference on Artificial Intelligence, vol. 33*. Washington, DC: AAAI Press, pp. 7378–7385. https://arxiv.org/abs/1811.05701.

Ye, Anbang, Christopher Cui, Taiwei Shi, and Mark O. Riedl (2023). "Neural Story Planning". In: *Proceedings of the AAAI-23 Workshop on Creative AI Across Modalities*. Washington, DC: AAAI Press. https://arxiv.org/pdf/2212.08718.

Contributions, Extensions and Future Directions   7

The best way to predict the future is to invent it.

—Alan Kay

7.1 Contributions

The summaries of each chapter are at the end of each, and need no repetition. Here I will highlight the contributions of this book.

Chapter 1 introduced storytelling, defining concepts like Narrative and Narrative Structure, using insights from the field of Narratology and literary studies. We then went on to discuss the fundamentals of Information Theory along with the basic techniques that underlie language models. This chapter is not the first time Narratology has been made relevant to computation and information theory, see Piper et al. (2021), but it is still, I think, unique in the accessible manner in which Narratology is tied together with linguistics-relevant issues of meaning, language universals, and biology.

Chapter 2 provided more or less standard fare introducing neural network architectures and some of their technicalities. The treatment is aimed at a broad audience, with emphasis on the intuitions behind the computational models. The reader who understands this chapter will have enough background to use existing tools with some insight, and to start reading the research papers in the field.

Chapter 3 introduced narratology's origins and central concepts. This is a far more in-depth explanation of narratological concepts than found elsewhere in computational accounts, along with the very first LLM-based experiments on focalization and a new theory of audience satisfaction through character evaluations. These elements are brought together in the annotation scheme called NarrativeML.

Chapter 4 provided a detailed look at plot as the central narrative concept, despite the lack of a consensus on a representation for plot. Classical narratological views of plot were discussed, particularly causal definitions, along with computational models and contemporary work. Comparisons among models highlighted computational limitations and implications for narratology in humanities-based studies. To support plot formalization, a plot-related layer was added to NarrativeML, involving character goals and Narrative Event Chains. This is not dramatically new material, but the presentation offers several linguistic and practical insights.

Chapter 5 covered time, a key concept in narrative. Some of the material will be familiar to the computational community using benchmarks involving temporal information extraction. Subordination relations are emphasized here, while they are ignored in many temporal tagging shared tasks, and so are habitual expressions. The approach to narrative pacing or tempo is new. The temporal features of NarrativeML allow for a richer characterization of narrative, whether for use in automatic analysis or generation.

Chapter 6 reviewed the evolution of story generation. The historical approaches may be familiar to many readers. However, the assessment of emerging neural trends and evaluation metrics is new, as is the assessment of the state of the art in story generation. The chapter proposed novel evaluation features for narrative generation: focalization, characterization, tempo, narrative distance, order, closure, suspense, and surprisal. Whether these will help overcome the weakness of evaluation metrics in recognizing diversity remains to be seen. Next, plot-related logic in terms of preconditions and postconditions were added to NarrativeML. Automatic analyses of micro-fictions using NarrativeML was shown to be well within the capabilities of LLMs, including extraction of plot-related information in first-order logic. Crucially, NarrativeML is expressive enough to use in prompts for story generation with LLMs, whether at the level of a plot outline or specifying the actual events. Finally, the chapter also stresses the importance of treating the code produced the story output as a first-class object of literary analysis, in keeping with Critical Code studies.

7.2 Extensions

I now discuss several extensions to the basic framework we have here, allowing us to specify spatial relations and motion. This naturally leads to automatic analysis of images by neural image processing with the output expressed by LLMs in NarrativeML.

7.2.1 Spatial Relations

7.2.1.1 Introduction

Natural language, as we discussed in Chap. 1, aims for efficiency, and it does so by allowing for imprecision, ambiguity and vagueness in its descriptions. Consider a simple example of route finding:

Example 7.1
The conference room is on the 22nd floor. Turn left immediately after the elevators and it's in the far left corner.

Though we can drop down when needed to technical language, including providing precise coordinates for a location, we usually describe objects in terms of qualitative relations between a 'figure' object F and a 'ground' object G. The relations can be:

- metric (i.e., distance and size) relations, such as 'near', 'adjacent to', 'far', 'three miles away', 'larger', and 'fainter'.
- topological relations, involving coincidence, contact, or containment, e.g., the prepositions 'in', 'on', 'over', and 'around');
- orientation (or direction) relations, involving a frame of reference that expresses relations between F, G, and in addition, possibly, a viewer V. Examples include 'over', 'above', 'east', 'around', 'to the left', and 'in front of'.

Understanding where a character is physically, and what they can see, is of considerable importance to judging the plausibility of event sequences for a computer-generated story, as well as in inferences focalization (as discussed in relation to focalized perceptions, in Chap. 3, Sect. 3.2.4.5).

The book by Mani and Pustejovsky (2012) builds upon the semantic distinctions underlying the annotation scheme SpatialML (Mani et al. 2010), which marks up geographic entities expressed by proper names as well as other noun phrases (facilitating geo-coding of text), as well as relations between places. It describes an extensive semantics for NL spatial relations, mapping them to calculi for qualitative spatial reasoning.

The next section goes into more technical detail, with some mathematics, and can be skipped if desired

7.2.1.2 Technical Details*

Just as temporal relations in NL can be mapped to the interval calculus, as shown in Chap. 5, topological relations can be mapped to the Region Connection Calculus (RCC-8) of Randell et al. (1992), which defines relations between 2D regions, in terms of whether they are disconnected, connected at their boundaries, overlapping, or one contained within the other, either touching the other's boundary or completely within it.

Orientation relations can rely on three different types of coordinate systems (Levinson 2003):

- In the *intrinsic frame* of reference, found in most languages, the coordinates are provided by particular facets of G, e.g., 'front', 'nose', 'sides', etc. To quote (Levinson 2003, pp. 42–43), "F lies in a search domain extending from G on the basis of an angle or line projected from the center of G, through an anchor point A (usually the named facet), outwards for a determined distance".

- The *absolute frame* of reference, which happens to be the only frame of reference for some languages such as the Australian language Guugu Yimithirr, involves a coordinate system that is anchored to fixed bearings with the origin on G, e.g., 'north of', 'uphill/downhill', 'towards the mountain/sea'.
- The *relative frame* of reference, which is absent in a third of the world's languages, involves a ternary relation, between F, G, and the viewer V; e.g., "the ball is to the left of the tree". There is one coordinate system centered on V, and possibly another, centered, as in the other two cases, on G. Naturally, this can result in considerable ambiguity, viz., "to the left: do you mean the left of the garage, or your left?"

Orientation relations are mapped to a number of different calculi. For the intrinsic frame, the Dipole Calculus (DyllaMoratz 2004) is used; for the absolute frame, the Cardinal Direction Calculus (GoyalEgenhofer 2000, Skiadopoulos 2005); and for the relative frame, the Double Cross Calculus (Freksa 1992, ScivosNebel 2001).[1]

As with the interval calculus used for time, these calculi too are intractable when the full set of qualitative relations are used, and thus they have to be restricted in various ways. To make it even more challenging, a given spatial expression may require a mapping to multiple calculi, e.g., "he swam around the pool" would require a combination of RCC-8 and the Dipole Calculus. Though not committing to any particular spatial calculi, NarrativeML has been extended (Mani 2016a) to accommodate such calculi, and has been used with all the calculi mentioned above.

Here are the spatial extensions to NarrativeML:

Example 7.2

```
<?xml version="1.0" encoding="UTF-8"?>
<!DOCTYPE NarrativeML [
...
<!ELEMENT NARRATIVE (#PCDATA | NARRATOR |
AUDIENCE | CHARACTER |
 EVENT | TIME | PLACE | TLINK | SLINK | SPATIALREL |
 EVALUATION | GOAL | TEMPO | NEC | PLOT|
 CONDITION | SEGMENT)*>
...
<!ELEMENT PLACE (#PCDATA)>
<!ATTLIST PLACE id ID #REQUIRED>
<!ATTLIST PLACE coref IDREF #IMPLIED>
```

---

[1] In the Double Cross Calculus, a line Y from point V to point G to ground is extended to create a pair of half-planes, left (l) and right (r). Two lines, one ($X1$) perpendicular to Y and through V, and the other parallel through it and through G, creates three regions, forward (f), back (b), with a central region (c) in between. These three lines constitute the Double Cross.

7.2 Extensions

```
<!ATTLIST PLACE exists (true | false) #IMPLIED>
<!ATTLIST PLACE accessibleTo IDREFS
#IMPLIED>
...
<!ELEMENT SPATIALREL EMPTY>
<!ATTLIST SPATIALREL id ID #REQUIRED>
<!ATTLIST SPATIALREL eventID IDREF #IMPLIED>
<!ATTLIST SPATIALREL predicate CDATA #IMPLIED>
<!ATTLIST SPATIALREL args IDREFS #IMPLIED>
...
]>
```

Given texts as input, these spatial relations are often extracted quite readily by LLMs. However, without evaluating the use of NarrativeML in shared tasks on spatial reasoning from text, it's premature to assess LLM capabilities here at this point. Likewise for the use of these spatial relations in generating plausible narratives.

To accommodate motion, further details of which are described in Mani (2016a), there are additional tags added to NarrativeML, associating EVENTs with ACTIONs that have logical decompositions into the motion predicates described in that work, along with a characterization of the EFFORT involved in the motion.

7.2.2 Image Analysis

Let us consider a frame from Martin Scorsese's award-winning film *Killers of the Flower Moon*, to be found in https://tinyurl.com/KFlowerMoon. Example 7.3 shows a one-shot example where we request Claude 3.5 Sonnet to generate spatial relations for the picture. The predicate calculus relations are removed, to make it easier to read and understand.

Example 7.3

> **User Prompt**: Please describe the spatial relations between the entities in the picture. Pay close attention to the format of the answer, which requires two justifications, each one with a different calculus.
>
> Question: What is the spatial relation between the open-topped jeep and the white house?
>
> Answer: *The jeep is in front of the white house.*
>
> Justification 1: There is space between the jeep and the white house, indicating, in terms of the Region Connection Calculus (RCC-8), that they are Disconnected (the relation DC in RCC-8).
>
> Justification 2: Relative to the viewer, equated with the camera, the white house is behind the jeep.

> **Machine**: 1. Man and Woman: *The man is to the left of and facing the woman.*
> <u>Justification 1</u>: The man and woman are Disconnected (DC) as there is clear space between them with no physical contact.
> <u>Justification 2</u>: From the viewer's perspective, the woman is in the left-front (lf) sector relative to the man.
> 2. Man and Car: *The man is standing next to the car.*
> <u>Justification 1</u>: The man and car are Externally Connected (EC) as the man appears to be in contact with or very close to touching the car.
> <u>Justification 2</u>: From the viewer's perspective, the car is in the back (b) or left-back (lb) sector relative to the man. 3. Woman and Car: *The woman is standing in front of the car.*
> <u>Justification 1</u>: The woman and car are Disconnected (DC) as there is visible space between them.
> <u>Justification 2</u>: From the viewer's perspective, the car is in the back (b) or right-back (rb) sector relative to the woman.

Overall, the machine does a decent job, but it makes two errors on the DCC relations (details not shown) for the first two entity pairs. A number of experiments with this LLM as well as GPT-4o on dozens of images reveal that they often fail to correctly identify the viewer's perspective, unsure of where the camera is. The object boundary detection is also imperfect. This results in many errors using qualitative calculi for the relative frame. Preprocessing the image with bounding boxes and other features from my own pipeline doesn't seem to improve the inferences. Nevertheless, the tools fare reasonably well on these examples in the other two frames of reference. Near-term advances in image processing should enhance these capabilities even further.

The extension of NarrativeML discussed in Mani (2016a) includes representation for changes in spatial configuration involving logics for motion, along with representation of the manner of motion. These extensions, the paper shows, have been applied to the annotation of movie corpora—which is now being done automatically using Gen AI tools. A method for generating narratives of silent movies, using NarrativeML as an internal representation of movie content, is found in Mani (2025).

7.3 Future Directions

7.3.1 Accelerating Studies in the Humanities

As I have argued elsewhere in Mani (2014), computational approaches have largely imported narratological theories that emphasize formal and/or logical structure, mapping narratological constructs to computational representations and illuminating interactions among them.

7.3 Future Directions

The flow of influences has historically been from narratology to computation. It is reasonable to expect such a flow to continue; for example, more elaborate narratological theories of character psychology, perhaps coming out of cognitive narratology, may be of considerable interest to computational approaches. As for the largely one-way direction of the flow, it is worth singling out some notable exceptions: (i) scripts, which continue to arouse narratological interest (but are only of marginal interest in computational approaches today, with their closest equivalent being NECs); and (ii) Plot Units, which have been discussed and elaborated by narratologists including Ryan (1991). One reason for a sluggish flow in the opposite direction may be attributed to the fact, alluded to earlier in Chap. 3, of the narratologist's focus on narrative discourse rather than fabula, and on form rather than content.

Generative AI is likely to provide useful services to humanities researchers. Consider a literary theorist or student wishing to analyze structural aspects of Chekhov's short story *The Lady with the Pet Dog*, which we discussed in Chaps. 3 and 5. Here is my interaction with GPT-4o:

Example 7.4

> **User Prompt**: Generate the NarrativeML for Chekhov's *The Lady with the Pet Dog*. Examine the NECs, and find out the events where they intersect. Determine (by some algorithm you explain) the prominent intersections where something significant happens to the characters. Give me those intersections in temporal order in a short text summary with embedded XML as needed. Finally, add a sentence giving the time line with temporal durations.
> **Machine**: To determine which intersections are most significant, we'll consider intersections that:
> Involve multiple NECs that indicate shifts in emotions or decisions.
> Include goals that are achieved or transformed, showing changes in character motivations.
> Mark the evolution of the relationship between Gurov and Anna, as highlighted by mental and action events.
> Algorithm for Prominent Intersections:
> <u>Identify NEC Intersections</u>: Locate where multiple NECs overlap across events, particularly where both mental and action events are present. This often suggests internal (mental) and external (action) shifts.
> <u>Goal Analysis</u>: Check if a character's goal is achieved, modified, or becomes newly prominent within an event. This is significant if it coincides with an NEC intersection.
> <u>Progressive Intensity</u>: Give weight to intersections where the intensity of emotions

Event	Description	Character Goal	NECS Gurov	NECS Anna
e1	Gurov, in Yalta, notices Anna with her pet dog.	Initiate contact	✓	✓
e2	Gurov initiates conversation.	Connect with Anna	✓	✓
e3	Gurov and Anna spend time together.	Develop intimacy	✓	✓
e4	Anna feels conflicted.	Resolve emotional confusion		✓
e5	They enter a romantic affair.	Pursue love	✓	✓
e6	Anna returns to her husband.	Obey societal duty		✓
e7	Gurov feels an unexpected longing for Anna after she leaves.	Understand his feelings	✓	
e8	Gurov travels to Anna's town to see her.	Reunite with Anna	✓	
e9	They meet secretly in Moscow and rekindle their relationship.	Reconnect romantically	✓	✓
e10	Anna feels torn between duty and love.	Reconcile inner conflict		✓
e11	Gurov realizes he truly loves Anna.	Accept deep love	✓	
e12	They discuss their future, determined to be together.	Plan a shared future	✓	✓

These events mark a narrative progression reflecting the deepening of their bond, from fleeting attraction to a profound love:
- over 1 minute for their first encounter,
- 3 weeks for their affair,
- 3 days of separation longing,
- 3 months until their reunion,
- 3 hours for their secret meeting,
- an indefinite future for their commitment to be together.

Fig. 7.1 GPT-4o NEC trajectories in *The Lady with the Pet Dog*

increases, moving from curiosity (NEC1) to realization and commitment (NEC5). Single-paragraph summary:

For reasons of space, the LLM's output table is shown separately as Fig. 7.1.

It can be seen that the results, which are accurate, are insightful as well as useful, drastically shortening study and teaching time. Using NarrativeML, the narrative progression, including duration estimates, considerably enhances the pedagogical experience.

We already saw, in Chap. 3, how AI was used to mine databases of tens of thousands of novels to assess literary theories. For example, research by Bamman et al. (2014) was able to analyze a database of more than 15,000 novels to test 30 literary hypotheses, and Elson et al. (2010) automatically analyzed conversational social networks in 19th century novels to test the claim of the literary critic (Moretti 1999). Those works were carried out prior to the use of LLMs, which can further augment such studies with more detailed inferences. In future, one might look at the types of narrative ordering or tempos across stories, and ask whether fast-paced stories are an invention of the post-industrial era.

7.3 Future Directions

Another interesting possibility for humanities scholars is extending NarrativeML to represent other kinds of information. For example, Kafka's story *The Metamorphosis* is arguably one of the most influential long-form short stories ever. Example 7.5 extends NarrativeML to accommodate several new facets (see the 'startPoint' added to the NARRATIVE tag, and the STYLE and NARRATIVE_DEVICE tags below):

- the way the narrative begins with the transformation already occurring, starting in-medias-res;
- the fact that there is a disjunction between his bodily insect state and his human mind;
- the overall bureaucratic tone and deadpan register;
- Kafka's use of defamiliarization (a term from Shklovsky 1973 that we introduced in Chap. 3) to reflect on the human condition.

Example 7.5

```
<NARRATIVE id="i1" narrator="n1" level="0" startPoint="in_medias_res>
<NARRATOR id="n1" type="absent" perspective="INTERNALLY_FOCALIZED"
          timeRelation="SUBSEQUENT" distance="NARRATED"
          order="CHRONICLE" />
<CHARACTER id="c1" name="Gregor Samsa" type="animate" exists="true"
           attributes="human_insect, traveling_salesman,
           overly_anxious" />
<EVENT id="e1" participants="c1">waking_up</EVENT>
<CONDITION id="cond1" event="e1" type="pre" logic="Asleep(c1)" />
<CONDITION id="cond2" event="e1" type="post" logic="Awake(c1)" />
<EVENT id="e2" participants="c1">discovering_transformation</EVENT>
<CONDITION id="cond3" event="e2" type="pre" logic="Awake(c1)
    AND Transformed(c1)" />
<CONDITION id="cond4" event="e2" type="post" logic="Aware(c1,
    Transformed(c1, insect))" />
<EVENT id="e3" participants="c1">thinking_about_job</EVENT>
<CONDITION id="cond5" event="e3" type="pre" logic="Aware(c1,
    Transformed(c1, insect))" />
<CONDITION id="cond6" event="e3" type="post" logic="Reflecting(c1,
    Responsibility)" />
<GOAL id="g1" character="c1" leaf="true" events="e3">maintain_job</GOAL>
<GOAL id="g2" character="c1" leaf="false">provide_for_family</GOAL>
<TLINK id="t1" type="BEFORE" eventID="e1" relatedToEvent="e2" />
<TLINK id="t2" type="BEFORE" eventID="e2" relatedToEvent="e3" />
<EVALUATION id="ev1" eventID="e3" characterID="c1" audienceID="reader1"
            polarity="negative" value="1"
            comment="Reader may perceive Gregor's mundane response
            to a grotesque transformation as unsettling or ironic,
            heightening the surreal tone." />
 <STYLE tone="bureaucratic" register="deadpan" surrealism="high"
    immediacy="abrupt" />
```

```
<NARRATIVE_DEVICE id="nd1" type="DISJUNCTION"
    character="c1" logic="BodilyState(c1, insect) AND
    MentalState(c1, human)" />
<NARRATIVE_DEVICE id="nd2" type="DEFAMILIARIZATION"
    target="e2" strength="strong" />
</NARRATIVE>
```

Defamiliarization could be elaborated further, in terms of its relation to identity of the character $c1$ and his attributes. These are just some of the extensions that a humanities scholar might add in the course of their research.

In the far future, interacting with the audience, improved reading and visualization environments might be developed (in some cases, leveraging dedicated book-reading devices), along with various checklists of narratological concepts that consumers might find interesting (an analogue of today's book discussion points). Or else, a system might expose a variety of interaction points via plans that accommodate user interaction, or it might switch on an audience simulator.

7.3.2 Revolutionizing Co-authoring of Literature?

Today's Gen AI is often used for brainstorming, relying on the mining of deep relationships from training data. When using it for generating stories, the average user goes through trial and error with text prompts, without benefiting from the higher degree of narratological specification and control provided by NarrativeML. As we have shown, the latter can be specified by hand in story prompts or generated automatically for use in prompting. The main limitation so far has been in terms of machine incoherence developing over longer stories. This issue is of course critical and would need to be addressed before the revolution can get underway. Generating appropriate dialogue with LLMs is also something I haven't explored much, and judging from the typical tone and registers mirrored from training data, it may be a while before LLMs can be counted on for snappy and creative dialogue.

As we have shown in the book, today Gen AI can produce automatically annotated text snippets that match narratological criteria. In the future, dialogue too may be in the purview of such tools. Sometimes the retrieval results (especially those which pertain to the inferred beliefs and motives of agents) may offer a different interpretation from the user's; this can in turn provide healthy fodder for narratological discussion. As discussed in Chap. 6 (Sect. 6.4.1.4), a Story Craft system might be trained to help revise authors' stories.

For the revolution to be successful, however, I do think we need to get a handle on judgments, namely the NarrativeML character evaluations by the audience, as specified in Chap. 3. If we were to get moderately high inter-annotator agreement across largish corpora on such evaluations, as opposed to small studies such as Elson (2012), the 'secret sauce' of what makes for good writing would finally be available to machines. That's because the neural machinery could be trained to optimize for these high audience evaluation scores. The existence of narratology, and the existence and deployment of other aspects of NarrativeML, and the various schemes it subsumes, indicates that narrative structure certainly isn't subjective. Why should audience judgments, which is after all part of our 'hearer models', be otherwise? When that sauce gets used, some writers, not just today's programmers who happen to be writers, will undoubtedly take advantage of the wins from collaboration with, and suggestions from, machines that help generate or augment stories at a plot or at a more fine-grained level. Throughout, of course, the ethical issues around authorship and authenticity discussed in Chap. 1 have to be addressed.

7.3.3 Revolutionizing Co-authoring of Movies?

CGI has gained an increasing role in movie authoring, from special visual effects and creating artificial characters to impressive visual modifications of real actors (including aging and de-aging using deep fake technology). While there has been considerable attention given in research to creating lifelike characters, the true potential of AI-invented characters has yet to be realized. Mining character attributes, relationships, and actions over vast training data can perhaps provide suggestions for truly artificial characters. For example, here is a prompt that can create an idea for a new character grafted from earlier ones:

Example 7.6
Give me the attributes of the character Macbeth and the events he participates in, including his NECs, in NarrativeML. Do the same for Hamlet. Then create a new character Hambeth who is a synthesis of the two and give me his NarrativeML. Generate a two-para story from Hambeth's NarrativeML.

Let us amplify this sort of mechanical grafting to scale. In that case, entire new 'reruns' are possible, except that they are hybrids of earlier movies or series. The human author can then use such mishmashes as raw material to recycle what has worked in the past, (which is unfortunately a substantial commercially-oriented goal of the mainstream film industry). Are we at a point where the next entirely machine-authored sequel for a TV series such as *Star Trek* or *Breaking Bad* is at hand? I think not. Even an episode is way beyond the length limits for coherent story generation, so script writers may heave a sigh of relief. However, there is considerable activity on social media with short AI-generated gifs and videos. I expect in due course further inroads will be made in video co-authoring. There may only be special niches in the process of movie development and production where AI is most effective. It's worth bearing in mind that the same point I made earlier about audience evaluations and a 'secret sauce' applies to movies.

7.3.4 Detecting Disinformation and Conspiracy Narrative Networks

Unfortunately, the growth of present day social networks has been accompanied by an unprecedented surge in conspiracy theories, which constitute narrative fiction. The divisive impacts of such fictions on society as a whole are well-known. The groundbreaking research of Tangherlini et al. (2020) has mined the online data corresponding to the well-known (and still influential) conspiracy theory called Pizzagate to automatically extract the narrative structure, loosely inspired by the narrative structure of Greimas (2010) that we mentioned in passing in Chap. 4. Given each sentence of the conspiracy data, they use off-the-shelf NLP pipelines to extract relationship triples.

For example, given the sentence "Podesta$_{x1}$ used the restaurant, Comet Pizza$_{x2}$, to hide a ring$_{x3}$ for trafficking in children." the triples would be: '$x1$ used $x2$', '$x1$ hid $x3$', and '$x2$ hosted $x3$'. The entities and relations are aggregated together into a graph and heuristically weighted, with various graph mining algorithms being applied, including community detection (a community is a set of nodes that are more densely connected within the set than without). For conspiracy theories, entities from different domains like politics, business, and entertainment are connected together in narrative threads, while factual conspiracies (like the so-called Bridgegate scandal) do not have such a property. When one of those threads is cut, the conspiracy theory falls apart, though it can grow again.

The relevant point here is that narrative structure can, when combined with graph mining, reveal the key characteristics of conspiracy theories. One would expect that NarrativeML, which is comparatively much more detailed and expressive than such relationship triples, could be highly relevant to characterizing conspiracy theories—and therefore might help debunk false information.

7.4 Final NarrativeML

The latest version (v. 2.2) of NarrativeML is too long to list in a readable way here. It lives at: https://tinyurl.com/5akfxsvs,

NarrativeML is expected to evolve somewhat further, so updates are likely in the future. As a reminder, corpora auto-annotated with NarrativeML are available here: https://tinyurl.com/suproc.

One limitation that the reader may have noticed is the lack of inter-annotator reliability for NarrativeML as a whole. As discussed in Chap. 5, the temporal aspects of NarrativeML have extensive reliability studies, with high reliability on events, times, and subordination, and lower reliability on event ordering relations. We pointed out there that the latter problem was due to the high density of the annotation required (which results in users leaving out details), and also due to the fact that a pair of superficially different annotations may be formally equivalent even if they are annotated differently. Approaches such as Cassidy et al. (2014) have used inference rules (e.g., transitive closure over temporal relations) to bridge the gap. Such problems are inherited by NarrativeML as a whole, and I think is inherent to annotating narrative content. Currently, **inter-annotator reliability** for NarrativeML on the ROC corpus ranges from 0.68 to 0.76 Kappa, with especially high agreement on Characters and Goals. The Appendix discusses practical guidelines for annotation while taking into account these issues. It is possible that developing more narrative-related inference rules may help with computing reliability.

Note that although I have pitched a single overall narrative annotation scheme for pedagogical purposes, my goal is not for it to be necessarily widely adopted per se. Rather, I would prefer if my work with this scheme could suggest and stimulate far more interest in computational exploitation of the narratological distinctions that NarrativeML encapsulates. My hope for the book as a whole is to bring greater clarity and understanding of the current capabilities, scope and prospects for Gen AI and humans in creatively engaging with the structures underlying the stories that embellish our lives.

References

Bamman, David, Ted Underwood, and Noah A. Smith (2014). "A Bayesian Mixed Effects Model of Literary Character". In: *Proceedings of the 52nd Annual Meeting of the Association for Computational Linguistics*, pp. 370–379. URL: https://aclanthology.org/P14-1035/.

Cassidy, Taylor, Bill McDowell, Nathanael Chambers, and Steven Bethard (2014). "An Annotation Framework for Dense Event Ordering". In: *Proceedings of the 52nd Annual Meeting of the Association for Computational Linguistics,* pp. 501–506. URL: https://aclanthology.org/P14-2082.pdf.

Dylla, Frank and Reinhard Moratz (2004). "Empirical complexity issues of practical spatial reasoning about relative position". In: *Workshop on Spatial and 2004, Valencia, Spain, August 2004.* Temporal Reasoning, ECAI.

Elson, David K. (2012). "Modeling Narrative Discourse". PhD thesis. Department of Computer Science, Columbia University.

Elson, David K., Nicholas Dames, and Kathleen R. McKeown (2010). "Extracting social networks from literary fiction." In: *Proceedings of the 48th Annual Meeting of the Association for Computational Linguistics (ACL'2010)*. Uppsala, Sweden, pp. 138–147. URL: https://aclanthology.org/P10-1015/.

Freksa, Christian (1992). "Using orientation information for qualitative spatial reasoning". In: *Theories and methods of spatiotemporal reasoning in geographic space*. Ed. by A. U. Frank, I. Campari, and U. Formentini. Berlin: Springer, pp. 162–178.

Goyal, R. and M. J. Egenhofer (2000). "Consistent queries over cardinal directions across different levels of detail". In: *Proceedings of the 11th International Workshop on Database and Expert Systems Applications, 2000.*

Greimas, Algirdas Julien (1984). *Structural semantics: an attempt at a method*. Lincoln, Nebraska: University of Nebraska Press.

Levinson, S. C. (2003). *Space in Language and Cognition*. Cambridge, UK: Cambridge University Press.

Mani, Inderjeet (2014). "Computational narratology". In: *Handbook of Narratology*. Ed. by Peter Huhn, Jan Christoph Meister, John Pier, and Wolf Schmid. Berlin: De Gruyter, pp. 84–92.

Mani, Inderjeet (2016a). "Animation Motion in NarrativeML". In: *7th Workshop on Computational Models of Narrative (CMN 2016)*. Vol. 5. 3. URL: https://doi.org/10.4230/OASIcs.CMN.2016.3.

Mani, Inderjeet (2025). "Automatic Narration of Movies via NarrativeML". In: *Workshop on Computational Models of Narrative (CMN'2025)*. URL: https://tecfa.unige.ch/cmn25/program.html.

Mani, Inderjeet, Christine Doran, et al. (2010). "SpatialML: annotation scheme, resources and evaluation". In: *Language Resources and Evaluation* 44.3, pp. 263–280. URL: https://doi.org/10.1007/s10579-010-9121-0.

Mani, Inderjeet and James Pustejovsky (2012). *Interpreting Motion: Grounded Representations for Spatial Language*. New York: Oxford University Press.

Moretti, Franco (1999). *Atlas of the European Novel, 1800-1900*. London: Verso.

Piper, Andrew, Richard Jean So, and David Bamman (2021). "Narrative Theory for Computational Narrative Understanding". In: *Proceedings of the 2021 Conference on Empirical Methods in Natural Language Processing*, pp. 298–311. URL: https://aclanthology.org/2021.emnlp-main.26/.

Randell, D. A., Z. Cui, and A. G. Cohn (1992). "A Spatial Logic on Regions and Connection". In: *Proceedings of 3rd Int. Conf. on Knowledge Representation and Reasoning (KR'1992)*, pp. 165–176.

Ryan, Marie-Laure (1991). *Possible Worlds, Artificial Intelligence and Narrative Theory*. Bloomington: Indiana University Press.

Scivos, Alexander and Bernhard Nebel (2001). "Double-Crossing: Decidability and Computational Complexity of a Qualitative Calculus for Navigation". In: *Proceedings of COSIT-2001*. Berlin: Springer.

Shklovsky, Viktor (1973). "On the connection between devices of syuzhet construction and general stylistic devices". In: *Russian Formalism*. Ed. by S. Bann and J. E. Bowlt. Edinburgh: Scottish Academic Press. URL: https://doi.org/10.1093/jts/flr173.

Skiadopoulos, Spiros and Manolis Koubarakis (2005). "On the consistency of cardinal direction constraints". In: *Artificial Intelligence* 163, pp. 91–135.

Tangherlini, Timothy R., Shadi Shahsavari, Behnam Shahbazi, Ehsan Ebrahimzadeh, and Vwani Roychowdhury (2020). "An automated pipeline for the discovery of conspiracy and conspiracy theory narrative frameworks: Bridgegate, Pizzagate and storytelling on the web". In: *PLoS ONE* 15.6. https://doi.org/10.1371/journal.pone.0233879.

NarrativeML Annotation Guidelines

This self-contained appendix goes into brief guidelines for annotating narratives using the NarrativeML annotation scheme. It explains how to mark up relevant parts of a narrative including events, entities including characters, and their narrative situations, and how they connect to each other. A narrative involves events that happen in a certain order, so representing the times when events happen and in what order is crucial. A narrative also involves characters acting according to certain goals, or at least reacting to certain events. The guidelines discuss how to use NarrativeML to mark up characters' goals and also the conditions that must be met for an event to happen and the conditions that result from the event (called post-conditions). These conditions are represented in a logic called first-order predicate calculus, with the predicates being open-domain relations rather than a fixed set of keywords or primitives. The annotation scheme also captures where events happen and how entities are related to each other spatially at different points in time. Finally, the guidelines show to capture a reader's reactions to the narrative, including whether an event outcome is good or bad for a particular character, or whether an ending is happy or sad. Because it is so thorough, annotating a story by hand can take a lot of time, so usually people use computer tools to assist in the annotation.

A.1 Introduction

NarrativeML involves marking up (or annotating) all the relevant parts of a narrative, like events, characters, and their narrative situations, and how they connect to each other.

A narrative involves events that happen in a certain order, so representing the times when events happen and in what order (e.g., whether event A is before event B or is included in it) is crucial. In NarrativeML, events and their temporal relations are represented using TimeML, which in turn leverages primitive relations from an AI reasoning formalism called the temporal interval calculus (Allen 1983, 1984). It also uses Narrative Event Chains

(NECs) (Chambers 2011), which lays out the order of events for each character and indicates where their paths intersect.

A narrative also involves characters acting according to certain goals, or at least reacting to certain events. NarrativeML lets you mark up characters' goals and also the conditions that must be met for an event to happen (called pre-conditions) and the conditions that result from the event (called post-conditions). These conditions are represented in a logic called first-order predicate calculus, with the predicates being open-domain relations rather than a fixed set of keywords or primitives.

For describing where events happen and how entities are related to each other spatially at different points in time, NarrativeML uses geographical markup from SpatialML (Mani et al. 2010) and primitives from AI reasoning formalisms called qualitative spatial calculi.[1] These spatial relations are based on further work that models, using such reasoning formalisms, a wide variety of static and dynamic spatial relations in natural language (Mani 2016b).

There's even a way to include the assumed reader's reactions – whether an event outcome for a particular character is appreciated by the reader as good or bad or whether an ending is happy or sad – following discussions from AI and narratology research.

NarrativeML is a highly expressive annotation scheme that covers an extremely wide range of stories found in text as well as other media. It involves multiple layers of annotation and has been applied to numerous stories. However, even within a layer the annotation is extremely dense and is thus very time-consuming for a human to annotate without machine assistance. Because it is so thorough, annotating a story by hand can take a lot of time, so usually people use computer tools to assist in the annotation.

A.2   Examples

The full specification of NarrativeML in terms of which annotation tags are valid and how they are structured is described in something called a DTD (Document Type Definition), which is shown in:

https://tinyurl.com/5akfxsvs.

You may want skim that now for a quick look at how everything is laid out, and then refer to it again when exploring each example.

A.2.1   Biking Blog Extract (Opening Sentence)

A.2.1.1 Source (Text)

Here is a simple example to illustrate the basics of NarrativeML. It doesn't require much reasoning to understand it.

---

[1] These include RCC-8 (Randell et al. 1992) and the Double Cross Calculus (Freksa 1992, Scivos-Nebel 2001).

Example A.1
March 7, 2006. Leaving San Cristobal, I biked with Gregg and Brooks for one more day.

A.2.1.2 Narrative Summary Text
Before doing NarrativeML annotation of the given text, one must first summarize what the annotation will contain. This is best done by writing a text paragraph. **Rule I: Write a summary of the source first (in terms of the characters, events, identifying what happened and possibly why)**.

A sample summary is as follows (make a habit of italicizing the characters, i.e., entities involved in the events, and underlining the mentioned events):

Example A.2
This very short narrative is told in the first person in past tense, with the events in the story beginning on a particular date. The three characters – the narrator, Gregg, and Brooks – left San Cristobal and then all of them biked for the rest of the day.

Before doing the annotation, here is another rule: **Rule II: Make sure you at least cover what's in the summary.** We call it The Golden Rule because it has the highest priority among the rules. Of course, this begs the question of what you put in the summary, which of course means honoring **Rule I**.

A.2.1.3 Annotation
Figure A.1 shows the annotation of Examples A.1 and A.2, taken together.

A.2.1.4 Discussion
It can be seen that the narrator is present in the story, which is in the past tense. There are three entities mentioned: the narrator, Gregg, and Brooks, and two events: leaving and biking. The leaving occurs during the date mentioned and precedes the biking, which also occurs during the date mentioned, which in turn takes a day. While leaving the bikers are externally connected to San Cristobal, and having left, they are disconnected from it. Each biker participates in the same above event chronology. The story covers one day and there is an estimated time for reading. The story is a very fast read, and spans, one assumes, last a day.

A.2.1.5 Notes
- The annotator is positing that "one more day" actually spans all of March 7, 2006. This may not be correct, as bikers may not bike all night. Or it might mean one day beyond March 7, 2006. Likewise the story time could be one day, but it could be two days as well. The next rule of annotation is: **Rule III. When in doubt, leave it out!** Accordingly, I would have disagreed with the annotator on that point.

```
<NARRATIVE id="i1" level="0">
<NARRATOR id="n1" name="Narrator" person="1sg" exists="true" distance="NARRATED" type="present"
          order="CHRONICLE" timeRelation="SUBSEQUENT"/>

<TIME id="t1" value="03-07-2006">March 7, 2006</TIME>, <EVENT id="e1"
participants="c1 c2 c3">Leaving</EVENT> <PLACE id="p1" name="San_Cristobal" exists="true"
latLong="" accessible_to="c1 c2 c3">San Cristobal</PLACE>, <MENTION id="m1" ref="c1">I</MENTION>
<EVENT id="e2" participants="c1 c2 c3">biked</EVENT> with <MENTION id="m2" ref="c2">Gregg</MENTION>
and <MENTION id="m3" ref="c3">Brooks</MENTION> for <TIME id="t2" value="P1D">one more day</TIME>.

<CHARACTER id="c1" name="Narrator" type="animate" coref="n1" exists="true" accessible_to="c1 c2 c3" />
<CHARACTER id="c2" name="Gregg" type="animate" exists="true" accessible_to="c1 c2 c3" />
<CHARACTER id="c3" name="Brooks" type="animate" exists="true" accessible_to="c1 c2 c3" />

<TLINK id="tr1" type="INCLUDES" timeID="t1" relatedToEvent="e1" />
<TLINK id="tr2" type="BEFORE" eventID="e1" relatedToEvent="e2" />
<TLINK id="tr3" type="INCLUDES" timeID="t1" relatedToEvent="e2" />
<TLINK id="tr4" type="SIMULTANEOUS" timeID="t1" relatedToTime="t2" />

<SPATIALREL id="sr1" eventID="e1" predicate="RCC8_EC" args="c1 p1" />
<SPATIALREL id="sr2" eventID="e2" predicate="RCC8_DC" args="c1 p1" />
<SPATIALREL id="sr3" eventID="e1" predicate="RCC8_EC" args="c2 p1" />
<SPATIALREL id="sr4" eventID="e2" predicate="RCC8_DC" args="c2 p1" />
<SPATIALREL id="sr5" eventID="e1" predicate="RCC8_EC" args="c3 p1" />
<SPATIALREL id="sr6" eventID="e2" predicate="RCC8_DC" args="c3 p1" />

<NEC id="nec1" entity="c1" events="e1 e2" />
<NEC id="nec2" entity="c2" events="e1 e2" />
<NEC id="nec3" entity="c3" events="e1 e2" />

<TEMPO storyTime="P1D" readingLength="PT4S" />
</NARRATIVE>
```

Fig. A.1 Annotation of biking blog extract

- The annotator has included accessibility information. Since the story doesn't have characters imagining or having things in their mind, let alone stories within stories, accessibility is not needed here. The spatial relations are included and are correct, stating that the bikers are in contact with Cristobal but they aren't especially interesting in this story. The characters' goals and the pre- and post-conditions of events have also been left out, along with evaluations, and that is acceptable as nothing important related to goals is happening. Another rule of annotation is: **Rule IV. Include only what's important for understanding the story!**

- What one annotator includes or leaves out will result in lowered inter-annotator agreement, or reliability. In other words, the annotators will differ in what they include or leave out, and common elements that they include may be tagged somewhat differently. In the research community on temporal relation annotation from text (with hundreds of papers on the subject), annotators have struggled with low reliability (sometimes as little as 50% agreement). The problem is due to the density of the annotation required, and also due to the fact that annotations may be formally equivalent even if they are annotated differently. In one approach, TimeBank-DENSE (Cassidy et al. 2014), the human annotation of temporal relations between events is expanded by the machine using transitive closure, e.g., if event A is before event B which in turn is before event C, A is inferred by the machine as also before C. That way, if one annotator didn't include that last relation, and the other did, they would be judged as equivalent. However, this is still not enough to ensure high reliability. Another more maximalist approach, MATRES (Ning et al. 2018) uses very strict rules on what to annotate, which involves dropping lots of relations in order to achieve much higher reliability.

- Last but not least, humans as well as machines make errors. Syntax errors can be easily addressed through validation of the NarrativeML. This can be done with tools like *xmllint* and *lxml*, among others. Other errors need to be detected by hand, mostly.

A.2.2 Spaghetti Story

Our next example is a simple text story from the ROC five-sentence story corpus (Mostafazadeh et al. 2016). Though short, it requires some commonsense reasoning to figure out what's happening.

A.2.2.1 Source (Text)
Example A.3
Tina made spaghetti for her boyfriend. It took a lot of work, but she was very proud. Her boyfriend ate the whole plate and said it was good. Tina tried it herself, and realized it was disgusting. She was touched that he pretended it was good to spare her feelings.

A.2.2.2 Narrative Summary Text
A sample summary following **Rule I** is as follows (as always, events mentioned are underlined and characters are italicized):

Example A.4
The story is narrated in the past tense by an external narrator. It concerns Tina, who works hard to make a dish of spaghetti for her boyfriend, presumably out of love for him. She is proud of the result, which might mean she assumes it's tasty, though the narrator doesn't directly say that. Her boyfriend eats it all and says it's good. Tina must assume, as a reader would at this point, that he's showing appreciation. But then she tries it herself, and realizes it's awful. That contradicts what her boyfriend said, so she has to come up with an explanation about what was going on in her boyfriend's mind when he said it. She is (one assumes) of a charitable disposition and assumes that people who love each other avoid hurting one another. So she is touched, assuming that he pretended it's good in order not to hurt her.

Note that the summary is much longer than the text, in violation of normal summarization conventions. The reason is that the summary makes explicit the goals and reasoning of the characters, which is left implicit in the original text. Understanding a narrative of this sort of course requires that level of abstract reasoning.

A.2.2.3 Annotation
Figure A.2 shows the annotation of Examples A.3 and A.4:

```xml
<NARRATIVE id="i1" level="0">
<NARRATOR id="n1" name="Narrator" person="3sg" exists="true" distance="NARRATED" type="absent"
     order="CHRONICLE" timeRelation="SUBSEQUENT" perspective="INTERNALLY_FOCALIZED"
     speechTime="t1" />
<CHARACTER id="c1" name="Tina" type="animate" exists="true" accessible_to="c1 c3" />
<CHARACTER id="c2" name="spaghetti" type="inanimate" exists="true" accessible_to="c1 c3" />
<CHARACTER id="c3" name="boyfriend" type="animate" exists="true" accessible_to="c1 c3" />

<MENTION id="m1" ref="c1">Tina</MENTION> <EVENT id="e1" participants="c1 c2">made</EVENT>
<MENTION id="m2" ref="c2">spaghetti</MENTION> for her <MENTION id="m3" ref="c3">boyfriend</MENTION>.
<EVENT id="e2.1" coref="e1">It</EVENT> took a lot of work, but <MENTION id="m4" ref="c1">she</MENTION>
was very <EVENT id="e2" participants="c1">proud</EVENT>. Her <MENTION id="m5" ref="c3">boyfriend</MENTION>
<EVENT id="e3" participants="c3 c2">ate</EVENT> the whole <MENTION id="m6" ref="c2" coref="c2">plate</MENTION>
and <EVENT id="e4" participants="c3">said</EVENT> <MENTION id="m7" ref="c2" coref="c2">it</MENTION> was good.
<MENTION id="m8" ref="c1">Tina</MENTION> <EVENT id="e5" participants="c1 c2">tried</EVENT>
   <MENTION id="m9" ref="c2" coref="c2">it</MENTION> herself, and <EVENT id="e6" participants="c1">realized</EVENT>
   <MENTION id="m10" ref="c2" coref="c2">it</MENTION> was disgusting.
   <MENTION id="m11" ref="c1">She</MENTION> was <EVENT id="e7" participants="c1">touched</EVENT>
   that <MENTION id="m12" ref="c3" coref="c3">he</MENTION> <EVENT id="e8" participants="c3">pretended</EVENT>
    <MENTION id="m13" ref="c2" coref="c2">it</MENTION> was good to spare her feelings.

<CONDITION id="cond1" event="e1" type="pre" logic="Exists(x) (Ingredient(x) AND Has(c1, x))" />
<CONDITION id="cond2" event="e1" type="post" logic="Cooked(c2)" />
<CONDITION id="cond3" event="e2" type="pre" logic="Cooked(c2)" />
<CONDITION id="cond4" event="e2" type="post" logic="Proud(c1, e1)" />
<CONDITION id="cond5" event="e3" type="pre" logic="OnPlate(c2)" />
<CONDITION id="cond6" event="e3" type="post" logic="not(OnPlate(c2))" />
<CONDITION id="cond7" event="e4" type="pre" logic="Ate(c3, c2)" />
<CONDITION id="cond8" event="e4" type="post" logic="Believe(c3, Good(c2))" />
<CONDITION id="cond9" event="e5" type="pre" logic="not(Empty(c2))" />
<CONDITION id="cond10" event="e5" type="post" logic="Tasted(c1, c2)" />
<CONDITION id="cond11" event="e6" type="pre" logic="Tasted(c1, c2)" />
<CONDITION id="cond12" event="e6" type="post" logic="Aware(c1, not(Good(c2)))" />
<CONDITION id="cond13" event="e7" type="pre" logic="Complimented(c3, c2)" />
<CONDITION id="cond14" event="e7" type="post" logic="Touched(c1, Complimented(c3, c2))" />
<CONDITION id="cond15" event="e8" type="pre" logic="Aware(c3, not(Good(c2)))" />
<CONDITION id="cond16" event="e8" type="post" logic="Pretended(c3, Good(c2))" />

<TLINK id="tr1" type="BEFORE" eventID="e1" relatedToEvent="e2" />
<TLINK id="tr2" type="BEFORE" eventID="e2" relatedToEvent="e3" />
<TLINK id="tr3" type="BEFORE" eventID="e3" relatedToEvent="e4" />
<TLINK id="tr4" type="BEFORE" eventID="e4" relatedToEvent="e5" />
<TLINK id="tr5" type="BEFORE" eventID="e5" relatedToEvent="e6" />
<TLINK id="tr6" type="SIMULTANEOUS" eventID="e6" relatedToEvent="e7" />
<TLINK id="tr7" type="SIMULTANEOUS" eventID="e8" relatedToEvent="e4" />

<NEC id="nec1" entity="c1" events="e1 e2 e5 e6 e7" />
<NEC id="nec3" entity="c2" events="e1 e3 e5" />
<NEC id="nec2" entity="c3" events="e2 e3" />

<GOAL id="g1" parent="g2" character="c1" leaf="true" events="e1">cook</GOAL>
<GOAL id="g2" parent="" character="c1" leaf="false">show love</GOAL>
<GOAL id="g3" parent="g2" character="c1" leaf="true" events="e2">feel proud</GOAL>
<GOAL id="g4" parent="" character="c2" leaf="false">show appreciation</GOAL>
<GOAL id="g5" parent="g4" character="c2" leaf="true" events="e4">say</GOAL>

<TEMPO storyTime="PT2H30M" readingLength="PT45S" />

<EVALUATION id="ev1" eventID="e4" characterID="c3" audienceID="reader1"
      value="1" polarity="positive" />
<EVALUATION id="ev2" eventID="e6" characterID="c1" audienceID="reader1"
      value="1" polarity="negative" />
<EVALUATION id="ev3" eventID="e7" characterID="c1" audienceID="reader1"
      value="1" polarity="positive" />
</NARRATIVE>
```

Fig. A.2 Annotation of spaghetti story

A.2.2.4 Discussion

It can be seen that the story is internally focalized, namely the narrator sees what Tina knows, while having access to her thoughts. There are three entities mentioned: Tina, the boyfriend, and the spaghetti. Tina has a goal g2 of showing love, which has a subgoal g1 of cooking (event e1), after which she has a subgoal g3 of feeling proud (event e2). There are also events of the boyfriend eating the spaghetti (event e3), followed by his saying (event e4) it's good, motivated by his goal (g4) of showing appreciation, then her trying it (event e5), and realizing (event e6) that it's not good, and being touched (event e7). Each of these events have pre- and post-conditions, and the events are in chronological sequence except for the simultaneous events of realizing and being touched. Tina participates in a sequence of five events, the boyfriend in two, and the spaghetti in three (indicated by the NECs), with obvious intersections. As for evaluations, the reader appreciates the boyfriend"s compliment (e4), is upset at Tina's realizing (e6) that it doesn't taste good, and at the end of the story is happy that Tina is touched (e7). There are also estimated durations for the story as a whole and its reading.

A.2.2.5 Notes
- Feeling proud is annotated as a subgoal of showing love, which is clearly incorrect. In general, identifying underlying goals is a source of disagreement among narrative annotators (Elson 2012).
- The annotator has left out accessibility information. S/he has also left out spatial relations. This is fine for this example as none of that really matters, so here the annotator has followed **Rule IV** correctly.
- The annotator has included annotations of personal pronouns. Including them may make it clear how the inferences are connected, establishing for example that the person who was touched was Tina. But is it in violation of **Rule IV**? It's unclear. In the sort of annotation tasks we are concerned with here, another rule also applies: **Rule V: Less is Better!** The reason here is not for the sake of our eyes as much as those of the machine. Having a machine do more dense annotation is likely to generate more errors.
- Another annotator may have different evaluations of the story, resulting in lowered reliability. Previous research (Elson 2012) has shown that annotators are able to carry out evaluations with high agreement, but these were of simple moralistic stories from Aesop's Fables.
- Since the annotations grow rapidly with the size of the source, with longer sources, or with non-textual sources, or do deal with overlapping rather than non-nested structures, we have to rely on standoff rather than inline annotation. In standoff annotation, the tags are stored separately from the source, and the tags have character or time offsets indicating their position in the source.

A.2.3 Video Story

Our next example is a 10-second video story from the VideoQA corpus (Li et al. 2022).

A.2.3.1 Source (Video)
Example A.5
https://tinyurl.com/52m5f5r7.

A.2.3.2 Narrative Summary Text

As always, we do a narrative summary first. Since the source is video, we will use standoff annotation. To simplify the annotation, we will skip time offsets. (Skipping time offsets is also worth doing in case the machine uses a narrative summary of the video that it itself has generated. Ideally, however, the time offsets would be included.) Here pre- and post-conditions are included, and indicated in parentheses.

Example A.6
Two children, appearing left-to-right as person_2 and person_1, are <u>playing</u> in a room with an entertainment center. On the floor there are colored dominoes arranged on a track, with a vertical block that serves as a maze partway on the track. At the start, person_2 is <u>sitting</u> to the far left near the start of the track and person_1 is <u>sitting</u> behind the middle of the track. (They can only do so if they have already started playing, and after taking their seats, they remain seated.) As soon as the dominoes start moving on the track, person_1 is seen <u>slinking</u> across to the far right near the end of the track – after which he is seen no more. (The dominoes can only move if they were properly arranged beforehand. Once they begin moving, they are considered to be in motion. Person_1 can only slink once he has been seated. He then reaches that far-right position.) All the while, person_2 remains <u>sitting</u> (which is possible only if he was already seated so he continues sitting). When a domino is nearing the bottom of the vertical maze, person_2 is seen <u>stepping</u> his foot forward; the rest of person_2 is occluded. (He can only step once he has been sitting. He is then considered to have stepped forward.) The dominoes are now <u>toppling</u> from the bottom of the vertical block to the end of the track, with person_2 no longer visible (which can only happen if they were already in motion, and at that point they are toppling). Once the toppling <u>stops</u> at the end of the track, we find person_2 <u>standing</u> (only his foot is visible) there at the end of the track (which must follow the toppling action, and now the dominoes are fully stopped, which only takes place once he has already stepped forward).

(The video is somewhat puzzling in that person_2's journey across the track is missing. It makes one suspect that at the end person_1 has been mislabeled as person_2.)

A.2.3.3 Annotation
Figure A.3 shows the annotation of Examples A.5 and A.6:

A.2.3.4 Discussion
Given a video, it is of course somewhat arbitrary what description one may provide for it, unless we stick to describing each of the scenes in terms of their spatial, temporal, and narrative-related aspects. This is a video about children playing, with dominoes moving left-to-right (mostly, except for the vertical maze) across the track, and children moving in the same left-to-right direction at particular times. The annotation focuses on the sequence

```xml
<NARRATIVE id="i1" level="0">
<NARRATOR id="n1" type="absent" />

<CHARACTER id="c1" name="person_1" type="animate" exists="true" />
<CHARACTER id="c2" name="person_2" type="animate" exists="true" />
<CHARACTER id="c3" name="dominoes" type="inanimate" exists="true"
       comment="this represents a set"/>

<PLACE id="p1" name="track" />

<EVENT id="e1" participants="c1 c2">playing</EVENT>
<EVENT id="e2" participants="c1 c2">sitting </EVENT>
<CONDITION id="cond3" event="e2" type="pre"  logic="Playing(c1, c2)" />
<CONDITION id="cond4" event="e2" type="post" logic="Sitting(c1) AND Sitting(c2)" />
<EVENT id="e3" participants="c3">moving</EVENT>
<CONDITION id="cond5" event="e3" type="pre"  logic="DominoesArranged(c3)" />
 <CONDITION id="cond6" event="e3" type="post" logic="Moving(c3)" />
<EVENT id="e4" participants="c1">slinking</EVENT>
<CONDITION id="cond7" event="e4" type="pre"  logic="Sitting(c1)" />
 <CONDITION id="cond8" event="e4" type="post" logic="AtLocation(c1, 'far_right')" />
<EVENT id="e5" participants="c2">sitting</EVENT>
 <CONDITION id="cond9"  event="e5" type="pre"  logic="AtLocation(c2, 'track_left')" />
 <CONDITION id="cond10" event="e5" type="post" logic="Sitting(c2)" />
<EVENT id="e6" participants="c2">stepping</EVENT>
<CONDITION id="cond11" event="e6" type="pre"  logic="Sitting(c2)" />
  <CONDITION id="cond12" event="e6" type="post" logic="StepForward(c2)" />
<EVENT id="e7" participants="c3">toppling</EVENT>
<CONDITION id="cond13" event="e7" type="pre"  logic="Moving(c3)" />
<CONDITION id="cond14" event="e7" type="post" logic="Toppling(c3)" />
<EVENT id="e8" participants="c3">stopping</EVENT>
<CONDITION id="cond15" event="e8" type="pre"  logic="Toppling(c3)" />
<CONDITION id="cond16" event="e8" type="post" logic="Stopped(c3)" />
<EVENT id="e9" participants="c2">standing</EVENT>
<CONDITION id="cond17" event="e9" type="pre"  logic="NotVisible(c2)" />
<CONDITION id="cond18" event="e9" type="post" logic="Standing(c2)" />

<TLINK id="tr1" type="INCLUDES"    eventID="e1" relatedToEvent="e2" />
<TLINK id="tr2" type="SIMULTANEOUS" eventID="e3" relatedToEvent="e4" />
<TLINK id="tr3" type="INCLUDES"    eventID="e5" relatedToEvent="e3" />
<TLINK id="tr4" type="INCLUDES"    eventID="e5" relatedToEvent="e4" />
<TLINK id="tr5" type="BEFORE"      eventID="e5" relatedToEvent="e6" />
<TLINK id="tr6" type="BEFORE"      eventID="e6" relatedToEvent="e7" />
<TLINK id="tr7" type="BEFORE"      eventID="e7" relatedToEvent="e8" />
<TLINK id="tr8" type="SIMULTANEOUS" eventID="e8" relatedToEvent="e9" />

<SPATIALREL id="sr1" eventID="e1" predicate="DCC_rf" args="c1 c2"
       comment="person_1 is to the right front of person_2" />
<SPATIALREL id="sr2" eventID="e2" predicate="DCC_lp" args="c2 p1"
       comment="person_2 is to the left of the track"  />
<SPATIALREL id="sr3" eventID="e2" predicate="DCC_sf" args="c1 p1"
       comment="person_1 is behind the middle of the track" />
<SPATIALREL id="sr4" eventID="e4" predicate="DCC_rp" args="c1 p1"
       comment="person_1 is at the far right of the track" />
<SPATIALREL id="sr5" eventID="e9" predicate="DCC_rp" args="c2 p1"
       comment="person_2 is at the far right of the track" />
</NARRATIVE>
```

Fig. A.3 Annotation of video story

of events, highlighting their temporal ordering and the spatial positions of the two children during those events. The dominoes' spatial relations are harder to capture, since the entire set of dominoes is individuated rather than individual dominoes.

A.2.3.5 Notes

- The whole video is about children playing. So the event e1 also INCLUDES all other events through e9. This is an application of **Rule V**, as including 8 more TLINKS makes the annotation much more dense.
- The character goals, tempo and evaluation are left out, since they are irrelevant. This is an application of **Rule IV** and **Rule V**.
- The NECs are also left out. It is true that the latter can always be trivially computed from the events and their participants; however, they are often useful in stories to indicate where the characters' trajectories intersect. They may also be useful in question-answering in case we need to know what one character did after doing something else, or predicting what a character should do next. Adding them in is left as an easy exercise.

A.3 Rules Recap

- *Rule I: Write a summary of the source first (in terms of the characters, events, identifying what happened and possibly why).*
- *Rule II: (The Golden Rule) Make sure you at least cover what's in the summary.*
- *Rule III. When in doubt, leave it out!*
- *Rule IV. Include only what's important for understanding the story!*
- *Rule V: Less is Better!*

References

Allen, James F. (1983). "Maintaining knowledge about temporal intervals". In: *Communications of the ACM* 26.11, pp. 832–843. URL: https://doi.org/10.1145/182.358434.

Allen, James F. (1984). "Towards a general theory of action and time". In: *Artificial Intelligence* 23.2, pp. 123–154. URL: http://doi.org/10.1016/0004-3702(84)90008-0.

Cassidy, Taylor, Bill McDowell, Nathanael Chambers, and Steven Bethard (2014). "An Annotation Framework for Dense Event Ordering". In: *Proceedings of the 52nd Annual Meeting of the Association for Computational Linguistics,* pp. 501–506. URL: https://aclanthology.org/P14-2082.pdf.

Chambers, Nathanael (2011). "Inducing Event Schemas and their Participants from Unlabeled Text". PhD thesis. Department of Computer Science, Stanford University.

Elson, David K. (2012). "Modeling Narrative Discourse". PhD thesis. Department of Computer Science, Columbia University.

Freksa, Christian (1992). "Using orientation information for qualitative spatial reasoning". In: *Theories and methods of spatiotemporal reasoning in geographic space.* Ed. by A. U. Frank, I. Campari, and U. Formentini. Berlin: Springer, pp. 162–178.

Li, Jiangtong, Li Niu, and Liqing Zhang (2022). "From Representation to Reasoning: Towards both Evidence and Commonsense Reasoning for Video Question-Answering". In: *Proceedings of Computer Vision and Pattern Recognition (CPVR'22).* URL: https://doi.org/10.48550/arXiv.2205.14895.

Mani, Inderjeet (2016a). "Animation Motion in NarrativeML". In: *7th Workshop on Computational Models of Narrative (CMN 2016)*. Vol. 5. 3. URL: https://doi.org/10.4230/OASIcs.CMN.2016.3.

Mani, Inderjeet, Christine Doran, et al. (2010). "SpatialML: annotation scheme, resources and evaluation". In: *Language Resources and Evaluation* 44.3, pp. 263–280. URL: https://doi.org/10.1007/s10579-010-9121-0.

Mostafazadeh, Nasrin, Nathanael Chambers, et al. (2016). "A corpus and evaluation framework for deeper understanding of commonsense stories". In: *Proceedings of the North American Chapter of the Association for Computational Linguistics: Human Language Technologies (NAACL-HLT'2016)*, pp. 839–849. URL: https://aclanthology.org/N16-1098.pdf.

Ning, Qiang, Hao Wu, and Dan Roth (2018). "A Multi-Axis Annotation Scheme for Event Temporal Relations". In: *Proceedings of the 56th Annual Meeting of the Association for Computational Linguistics (ACL'2018)*. URL: https://aclanthology.org/P18-1122/.

Randell, D. A., Z. Cui, and A. G. Cohn (1992). "A Spatial Logic on Regions and Connection". In: *Proceedings of 3rd Int. Conf. on Knowledge Representation and Reasoning (KR'1992)*, pp. 165–176.

Scivos, Alexander and Bernhard Nebel (2001). "Double-Crossing: Decidability and Computational Complexity of a Qualitative Calculus for Navigation". In: *Proceedings of COSIT-2001*. Berlin: Springer.

Index

A
Aarseth, 193
A Beautiful Mind, 105, 107, 145
Abelson, 134, 135
Abrams, 126
Accessibility, 105
Accessibility set, 106
Achrony, 157, 163, 198
Activation function, 44
Actual reader, 111, 193
Ahn, 166
Allen, 158, 160
Anagnorisis, 126, 141
Analepsis, 157, 163, 198
Aristotle, 111, 124–126
Asher, 166
Audience, 6, 110, 111, 193
Audience response, 111
Austerlitz, 87, 154
Author, 87
Authorial audience, 111

B
Backprop, 58
Badia, 162
Bal, 6
Bamman, 109, 125, 232
Barthes, 170
Bartlett, 134
Basu, 76, 207

Bates, 194
Baumard, 1
Beat goal, 196
Belief revision, 105
Berglund, 167
Bernardo, 197
BERT, 77
Bertram, 189, 219
Bethard, 164, 172
Bias, 28
Bigrams, 10
Bittar, 161
Bod, 133
Boden, 33
Booth, 92, 105, 111
Borgeaud, 79, 204
Bosselut, 204
Bramsen, 167
Branco, 161
Braun, 190
Brewer, 145
Brown, 15
Bullet time, 175
Bunt, 161
Burckert, 160

C
Campbell, 127
Candide, 175
Case-Based Reasoning, 190, 192, 193

Caselli, 161
Cassidy, 165
Chambers, 142
Character, 108, 109, 172
Character goals, 189
Chatman, 6, 174
Chekhov, 231
Cheong, 197
Chinese Room, 24
Chronicle, 157, 163, 198
Churning, 191
Cinderella, 6
Cloze, 19
Colton, 33
Compactness, 125
Complication, 126
Computational narratology, 86
Cortazar, 103
Cost, 82
Costa, 161
Costs, 30
CoT prompting, 80
Cross-entropy, 15
Cross-entropy loss, 57, 62
Cullingford, 134
Cyropedia, 87

D
Days of Our Lives, 191
Declerck, 131, 191
DeepSeek, 79
Defamiliarization, 233
DeJong, 135
Denotational Semantics, 21
De Saussure, 22
Diegesis, 89
Diegetic level, 99
Direct speech, 90
Discours, 6
Discourse, 6
Discourse analysis, 4
Discourse time, 172, 174, 177
Distributional Semantics, 22, 60
Donatelli, 26
Doxastic preference, 139
DramaBank, 141
Dropout, 57, 60
Dubourg, 1

Dundes, 129
Duration, 174
Dylla, 228

E
Ebell, 31
Edmiston, 92
Egenhofer, 228
Eldan, 30
Elson, 109, 114, 140, 141, 172, 235
Embassytown, 12
Embedding norm, 106
Entropy, 13
Episode, 125
Evaluating event outcomes, 112
Evaluation, 206
Event time, 154
Existence, 106
Externally-focalized perspective, 91
Extradiegetic level, 99

F
Fabula, 6
Factuality profile, 118, 180
Fan, 36, 203
Fatemi, 165
Feed Forward Neural Net, 51
Femme fatale, 111
Ferro, 160, 161
Fine-tuning, 78
Finlayson, 132, 133
Finnegan, 190
First-person narrative, 87
Firth, 22
Fitzgerald, 89
Fluency, 18
Focalization, 91, 96
Focalization with LLMs, 93
Focalized perceptions, 98
Focalizer, 91
Forascu, 161
Forster, 5, 6, 124
Forward Pass, 55
Frame story, 99
Frank, 161
Free indirect style, 90
Freksa, 228

Freytag, 126
Futrell, 26

G
Genette, 6, 87, 89–91, 99, 104, 111, 157, 163, 172, 175, 177, 190
Gerrig, 197
Gervas, 133, 190, 193
Gibson, 26
Gillespie, 219
GloVe, 62
Goyal, 138, 228
GPT embeddings, 63
Graesser, 157
Grall, 157
Grasbon, 190
Greenberg, 26
Greimas, 133, 236

H
Habituals, 154, 171
Hahn, 26
Hallucinations, 28
Halpern, 106
Hamlet, 126
Harris, 22
Hawaii, 175
Hayes, 160
He, 25
Heart of Darkness, 87
Herman, 5, 86
Heroic quest, 127
Heterodiegetic narrator, 87
Histoire, 6
Hobbs, 106
Hogan, 112
Homodiegetic narrator, 87
Horace, 126
Howe, 26
Huang, 168
Hwang, 204

I
Iliad, 89, 124
Imitation, 89
Immediate speech, 90
Implied author, 111

Implied reader, 111
Incoherence, 75
In medias res, 126
Information content, 11
Information theory, 7
Interior monologue, 90
Internally-focalized perspective, 91
Intradiegetic level, 99
Ippolito, 203
Ireland, 175
Irrealis particle, 152
Iser, 111
ISO-Space, 118
Iyyer, 109

J
Jahn, 175
Jealousy, 157, 159, 175
Jean Santeuil, 104, 157, 163
Jin, 92
Jockers, 5
Jones, 144
Jurafsky, 42, 49

K
Kafka, 233
Kaplan, 82
Kappa, 112
Kautz, 159, 186, 188
Kellogg, 87
Kelly, 204
Killers of the Flower Moon, 230
KL divergence, 16
Knowledge norm, 106
Koller, 26
Kolodner, 190
Kolomiyets, 172
Koubarakis, 228
Kripke, 105
Kukkonen, 145

L
Ladkin, 159
Language Modeling Head, 74
Language universals, 26
Lascarides, 166
Laurel, 127

Lebowitz, 191
Lehnert, 135, 137, 138, 140, 142
Les Liaisons Dangereuses, 88
Levinson, 228
Levshina, 26
Lewis, 79
Li, 30, 167, 202
Lichtenstein, 145
Life: A User's Manual, 157
Lin, 152
Liu, 92
LLM, 19
Lonneker, 197
Lord, 190
Loss, 46, 57, 62
Lowe, 139
Loyall, 194
Ludology, 193

M
Mahabharata, 87
Mamba, 82
Mani, 27, 35, 86, 154, 157, 160, 164, 167, 174, 177, 179, 180, 228, 231
Manning, 25
Marino, 220
Martin, 6, 42, 49
Masking, 69
Mateas, 127, 194, 195
McAfee, 25
McHale, 103
McKee, 196
McLelland, 81
Meaning, 9
Meehan, 189
Meister, 85, 86, 177
Mendez, 133
Metadiegetic level, 99
Metalepsis, 102, 145, 171
Metametadiegetic level, 99
Mieville, 12
Mikolov, 60, 61
Mimesis, 89
Mini-batch, 46
Mirroring, 18
Monomyth, 128
Montfort, 88, 189, 198, 219, 220
Moratz, 228

Moretti, 109, 232
Moses, 106
Mostafazadeh, 4, 24, 34, 146
Mr. Palomar, 124
Mrs. Dalloway, 90
Multi-head Attention, 69
Murray, 127, 193
My Apprenticeship, 87
Mythos, 124

N
Narrated speech, 90
Narratee, 111
Narrateme, 129, 130
Narrative, 4
Narrative arc, 125
Narrative distance, 89
Narrative Event Chain, 142
Narrative function, 129
Narrative levels, 99
NarrativeML, 115, 146, 177, 236
Narrative ordering rule, 166
Narrative structure, 5
Narratology, 85, 86
Narrator, 6
Narrator identity, 87
Narrator perspective, 91
Nebel, 160, 228
Necessity, 105
N-grams, 10
Nieto, 161
1984, 115, 145
Ning, 165
Non-focalized perspective, 91
Nucleus sampling, 75

O
Odyssey, 125
Oliver Twist, 112
One Hundred Years of Solitude, 162
Ontological promiscuity, 106
Opacity, 28, 68, 70
Ordering, 157
Outcome, 194
Overbeeke, 161
Overfitting, 57, 60

P

Pace, 175
Pan, 174, 176, 177, 179
Panchatantra, 87
Pang, 115
Paralepsis, 92
Paralipsis, 91
Parsons, 106, 107
Path consistency, 160
Peinado, 190
Peng, 203, 204
Pennington, 62
Perceptron, 42
Perez, 193
Peripeteia, 126, 141
Perplexity, 20
Phelan, 115
Piantadosi, 9
Pier, 103
Pillutla, 207
Pinker, 24
Piper, 119, 225
Pity, 111
Planning, 186
Plans, 186
Plato, 89
Plot, 7, 123, 124, 172, 186, 202, 214
Plot unit, 135, 137
Polysemy, 21
Possibility, 105
Possible worlds, 105
Post-classical narratology, 86
Prince, 111
Prior time relation, 155
Prolepsis, 157, 163, 198
Prompts, 79
Propp, 129
Pustejovsky, 27, 104, 161, 166, 179, 180, 228

Q

Quinn, 102, 207
Quotation, 90

R

Rabinowitz, 111, 112
Ramanujan, 133
Reactive planning, 194
Reader, 110
Reader affect, 112
Realis particle, 152
Reference time, 154, 198
Reichenbach, 154
RelU, 49
Repetition, 75
Replanning, 194
Reported speech, 90
Retrieval Augmented Generation (RAG), 79
Retrograde, 157, 163, 198
Riedel, 187, 193
Riedl, 29, 200, 210
ROC corpus, 36, 93, 96, 97, 116, 153, 156, 199, 202, 208, 214, 216
Rockwell, 5
Rosenblatt, 42
Ryan, 5, 154, 231

S

Safety, 31
Sahlgren, 22
Sarrasine, 170, 171
Sauri, 104, 161, 166, 180
Schank, 134, 135
Scheidel, 131, 191
Schema, 134
Scholes, 87
Scivvos, 228
Scripts, 134
Searle, 24
Self-attention, 65, 66, 69
Selman, 188
Sentence-probability, 17
Shannon, 7
Sharples, 193
Shipley, 25
Shklovsky, 6, 103, 233
Showing, 89
Sigmoid function, 49
Simple narration, 89
Simultaneous time relation, 155
Sinclair, 5
Sjuzhet, 6
Sketchy script, 135
Skiadopoulos, 228
Smith, 1, 125, 205
SpatialML, 118

Speculative language, 104
Speech time, 154, 198
Spreyer, 161
Stative rule, 166
Step function, 44
Stern, 127, 194, 195
Sternberg, 145
Sterne, 103
Stochastic Gradient Descent, 47, 58
Story, 4, 6, 124
Story generation, 186
Story Intention Graph, 140
Story time, 172, 175, 177
Subordinated discourse, 104
Subsequent time relation, 155
Surprisal, 14, 76
Syllepsis, 157, 163, 198

T
Tangherlini, 236
Taylor, 19
Telling, 89
Temperature, 75
Tempo, 172, 175, 177, 179
Temporal graph, 160
Tess of the d'Urbervilles, 115, 145
Thalken, 5
The Catcher in the Rye, 110
The Continuity of Parks, 103, 107, 145
The Dead, 92
The Fox and the Crow, 144
The Genius, 92
The Golden Ass, 87
The Hunger Artist, 154
The Killers, 88, 91
The Lady and Five Suitors, 99
The Lady with the Pet Dog, 113, 159
The Magic Swan Geese, 131, 190
The Murder of Roger Ackroyd, 105
The Short Happy Life of Francis Macomber, 91
The Sound and the Fury, 101
The Tale of the King, 99
The Thousand and One Nights, 87, 99, 106, 145
The Three Ladies of Baghdad, 99
The Travelers and the Bear, 172

The Waves, 88, 124
The Wily Lion, 129, 140, 172
Time's Arrow, 157
TimeBank, 161, 165
TimeML, 118, 161, 162, 164
TIMEX2, 160
Todorov, 91
Toolan, 5
Top-k decoding, 75
Transformer, 70
Transformer specs, 77
Tristram Shandy, 110, 191
Truth, 28
Tuckute, 25
Turing, 24
Turner, 193

U
Underwood, 174
Universal approximation theorem, 54
Unraveling, 126, 127
Unreliable narrator, 105
User model, 111, 196

V
Van Andel, 140
Vanishing gradient, 57, 60
Vaswani, 65
Venkatraman, 32
Verhagen, 166
Vilain, 160

W
Wallace, 101
Wang, 25
Wardrip-Fruin, 189, 192, 193
Weaver, 7
Webber, 166
Wei, 79, 80
Who's Afraid of Virginia Woolf, 194
Wiggins, 33
Wild Palms, 125
Word embeddings, 60
Word2vec, 60, 61
Wright, 31

X
Xie, 200, 210
XOR, 49

Y
Yao, 203
Yarlott, 133
Yauney, 174
Ye, 202

Yoshikawa, 168
Young, 187, 193, 197

Z
Zero-focalized perspective, 91
Zigzag, 157, 163, 198
Zipf, 9
Zwaan, 157

GPSR Compliance

The European Union's (EU) General Product Safety Regulation (GPSR) is a set of rules that requires consumer products to be safe and our obligations to ensure this.

If you have any concerns about our products, you can contact us on

ProductSafety@springernature.com

In case Publisher is established outside the EU, the EU authorized representative is:

Springer Nature Customer Service Center GmbH
Europaplatz 3
69115 Heidelberg, Germany